T0360016

MATHS SEQUENCES
FOR THE EARLY YEARS

CHALLENGING CHILDREN TO REASON MATHEMATICALLY

PETER SULLIVAN

JANETTE BOBIS

ANN DOWNTON

SHARYN LIVY

MELODY McCORMICK

JAMES RUSSO

OXFORD
UNIVERSITY PRESS

OXFORD
UNIVERSITY PRESS

Oxford University Press is a department of the University of Oxford.
It furthers the University's objective of excellence in research,
scholarship, and education by publishing worldwide. Oxford is a registered
trademark of Oxford University Press in the UK and in certain other countries.

Published in Australia by
Oxford University Press
Level 8, 737 Bourke Street, Docklands, Victoria 3008, Australia.

© Peter Sullivan, Janette Bobis, Ann Downton, Sharyn Livy, Melody McCormick and James Russo 2023

The moral rights of the authors have been asserted

First published 2023
Reprinted 2023 (twice)

All rights reserved. No part of this publication may be reproduced, stored in a retrieval system, or
transmitted, in any form or by any means, without the prior permission in writing of Oxford University
Press, or as expressly permitted by law, by licence, or under terms agreed with the reprographics rights
organisation. Enquiries concerning reproduction outside the scope of the above should be sent to the
Rights Department, Oxford University Press, at the address above.

You must not circulate this work in any other form and you must impose this same condition on
any acquirer.

 A catalogue record for this
book is available from the
National Library of Australia

ISBN 9780190338541

Reproduction and communication for educational purposes
The Australian *Copyright Act 1968* (the Act) allows educational institutions that
are covered by remuneration arrangements with Copyright Agency to reproduce
and communicate certain material for educational purposes. For more information,
see copyright.com.au.

Edited by Vanessa Lanaway
Typeset by Newgen KnowledgeWorks Pvt. Ltd., Chennai, India
Proofread by Tom Smallman
Printed in China by Leo Paper Products Ltd

Disclaimer
*Links to third party websites are provided by Oxford in good faith and for information only. Oxford disclaims
any responsibility for the materials contained in any third party website referenced in this work.*

CONTENTS

ACKNOWLEDGEMENTS

This book is one of the outcomes of the *Exploring Mathematical Sequences of Connected, Cumulative and Challenging Tasks* (EMC³) research project funded by the Australian Research Council, Catholic Education Diocese of Parramatta (CEDP) and Melbourne Archdiocese Catholic Schools (MACS). The collaboration was led by Peter Sullivan and academic colleagues Ann Downton, Sharyn Livy, James Russo (Monash University) and Janette Bobis (University of Sydney). Other colleagues who supported the project were Melody McCormick, Sally Hughes, Jane Hubbard, Maggie Feng and Ellen Corovic.

Implementation of the project and development of resource materials were supported by our system partners Paul Stenning and Barbara McHugh (CEDP), James Giannopoulos (MACS) and teachers in the project who trialled the lessons in their classrooms. The project was also supported by the Teaching Educators from CEDP and Mathematics Learning Consultants from MACS.

We acknowledge and thank teachers from Foundation Level to Year 2 in the following schools for their support in trialling the sequences of lessons and providing feedback:

Victorian primary schools

Cheltenham Primary, Deer Park Primary, Eaglehawk Primary (Bendigo), Elwood Primary, Gilson College, Hampton Primary, Lightning Reef Primary (Bendigo), Middle Park Primary, Patterson Lakes Primary, School of Good Shepherd (Gladstone Park), Specimen Hill Primary (Bendigo), St Anthony's (Alphington), St Augustine's (Keilor), St Bernard's (Wangaratta), St Fidelis (Moreland), St Francis de Sales (Lynbrook), St John's XXIII (Thomastown East), St Joan of Arc (Brighton), St John's (Mitcham), St Joseph's (Brunswick West), St Kilda Primary, Saint Margaret's Primary School (Maribyrnong), St Martin de Porres (Laverton), St Martin of Tours (Rosanna), St Mary of the Cross Mackillop (Epping North), St Michael's (North Melbourne), St Monica's (Moonee Ponds), St Peter's (Bentleigh East), Star of the Sea (Ocean Grove), St Therese's (Essendon), Tulliallan Primary (Cranbourne North), Violet Street (Bendigo), Warrnambool East Primary.

NSW primary schools

Chisholm (Bligh Park), Christ the King (North Rocks), Corpus Christi (Cranebrook), Holy Cross (Glenwood), Holy Spirit (Lavington), Holy Spirit (St Clair), Holy Trinity (Granville), Mother Teresa (Westmead), Our Lady of Lourdes (Seven Hills), Our Lady of the Rosary (St Mary's), Our Lady of the Way (Emu Plains), Sacred Heart (Mount Druitt), St Aidan's (Rooty Hill), St Anthony's (Girraween), St Bernadette's (Castle Hill), St Bernadette's (Lalor Park), St Canice's (Katoomba), St Finbar's (Glenbrook), St Francis of Assisi (Glendenning), St Joseph's (Schofields), St Madeleine's (Kenthurst), St Mary's (Rydalmere), St Michael's (Blacktown South), St Oliver's (Harris Park), St Patrick's (Blacktown), St Patrick's (Parramatta), St Thomas Aquinas (Springwood), Trinity (Kemps Creek).

We are also grateful to Colleen Vale (Monash University) and Jenni Way (University of Sydney), members of the Advisory Board, for their support and insights throughout the project.

INTRODUCTION

This book presents 14 sequences of learning experiences, arranged according to the key domains of the curriculum, to support the teaching of mathematics in the early school years (students aged 5 to 8). Each sequence is made up of a series of 'suggestions' that build on one another and support the learning of an aspect of the respective sequence. The following specific information for teachers is embedded within each sequence:

- a summary of the suggestions including a statement of mathematical focus
- an explanation of the rationale for the sequence including big ideas and elaboration of learning goals
- a statement of pre-requisite and new language
- ideas for launching the learning experiences
- specific statements of the intended learning that can inform assessment.

We consider the ideal curriculum to be made up of descriptions of content (nouns, if you like) **and** proficiencies (understanding, fluency, problem solving and reasoning) that can be thought of as verbs. The resources presented in this book are based on the posing of *problems* allowing for student *reasoning* that in turn build relational *understanding* and foster *fluency*. This approach is termed teaching *through* problem solving (Schroeder & Lester, 1989).

The sequences provide opportunities for students to seek insights for themselves within a consistent lesson structure. The approach we have taken challenges the assumption that the optimal way to teach mathematics is by explicitly telling students what to do, followed by practice. In this book, explicit teaching occurs **after** the learning *through* problem solving, so that students have had opportunities to engage with the problem or task prior to the central ideas being explained.

Posing challenging tasks in which students engage prior to instruction, when supported by appropriate pedagogies, effectively engages most students in creating and learning mathematics. The pedagogies have all been trialled in classrooms and have the potential to enhance the planning, teaching and assessment of mathematics.

THE EARLY LEARNING OF NUMBER

To some extent, most of the resources presented in this book assume that students are already fluent with saying the number sequence forward to 10 or beyond and can allocate the correct number to quantify a collection. However, we realise that some students require additional support in achieving both of those goals.

Appendix 3, 'Early learning of number' (page 173), explores this in depth, providing numerous suggestions for teachers seeking ways to provide their students with additional support. The intention of these suggestions is that the whole class participates, even though the majority of the class may be fluent with saying the number sequence and counting collections. This is important, both to maximise learning opportunities for all students, and to ensure inclusion, which is one of our highest priorities for all children when learning mathematics.

The suggestions provided in 'Early learning of number' explore the following approaches in detail:

- developing familiarity with the number sequence (stable order principle)
- using numbers to quantify a collection (cardinality)
- counting once and only once (one-to-one correspondence)
- the quantity is independent of the way the objects are arranged (conservation).

By providing students with a range of learning experiences that focus on important ideas in early counting, in particular number sequences, using numbers to quantify a collection, counting using one-to-one correspondence and number conversation, they will have opportunities to develop their conceptual understanding in each of these areas with the aim of later participating in learning mathematics through challenging tasks without requiring significant enabling prompts or supports.

OVERCOMING OBSTACLES

Working on challenging tasks requires students to engage in sophisticated mathematical behaviour, including: planning how they will approach the task; choosing appropriate strategies and explaining these choices; recording their thinking and using this record to support mathematical reasoning. Equally important are the students' dispositions when engaged with such tasks. Ideally, students should take responsibility for their learning, deciding when they need support and when they are ready for extension. They should display a willingness to tolerate confusion, and to persevere when they are struggling.

Teachers in the early years of schooling, particularly Foundation teachers, are often uncertain about whether this type of mathematical work is appropriate for their students. The four main obstacles to working on challenging tasks identified by teachers are:

1 The belief that challenging tasks are not for all students
2 Not knowing when to begin teaching with challenging tasks
3 The feeling that some tasks are not meaningful for students (i.e. are too mathematically abstract)
4 The belief that students cannot cope with the struggle.

Appendix 2, 'Overcoming obstacles' (page 167), explores ways to overcome these challenges in detail, but the key unifying mechanism in overcoming these obstacles was a change in teachers' mindset. As teachers began to use these sequences, they overcame their initial paralysis, that is, 'not knowing when to begin teaching with challenging tasks' and the belief that 'students cannot cope with the struggle'. Moreover, the fact that most students seemed to benefit from exposure to these sequences and tasks helped to undermine the 'belief that challenging tasks are not for all students'. Finally, as these initial pedagogical experiences allowed teachers to gain further confidence, they became more comfortable modifying and adapting tasks, helping them to directly address their concerns 'that some tasks are not meaningful for students'.

OXFORD UNIVERSITY PRESS

CONCLUSION

In the research project within which these resources were developed, teachers who utilised the resources reported improved student engagement, better developed understanding, and more productive dispositions. From a teacher perspective, employing these pedagogies led to greater valuing of productive struggle (Russo et al., 2020), supported differentiated instruction, and fostered a growth mindset characterised by the belief that all students can learn mathematics (Bobis et al., 2021). Teachers also reported that planning time was simplified and reduced since the resources assume whole class mixed achievement student grouping.

We encourage teachers to read fully any sequences they are intending to teach and to imagine the trajectory of learning encapsulated by the resources. The pedagogies have all been trialled in classrooms and have the potential to enhance planning, teaching and assessment of mathematics.

Overview

This book contains 14 sequences of learning that collectively address most of the mathematics curriculum for the first three years of schooling. Each sequence represents an important macro mathematical concept that students should explore as part of their school mathematics program. There are five sequences in the number domain ('Counting principles'; 'Structure of number'; 'Making things equal'; 'Counting patterns'; and 'Place value'), four sequences in the measurement domain ('Informal length measuring'; 'Time'; 'Informal approaches to perimeter and area'; 'Volume'), four sequences in the geometry domain (Recognising polygons; Objects; Reasoning with polygons; Location and transformation), and one sequence in the probability and statistics domain.

Note that not all sequences are equally appropriate for all year levels. For example, we suggest that Year 2 teachers are less likely to use the 'Counting principles' sequence, whilst teachers of students in their first year of school are perhaps less likely to explore the 'Place value' sequence (see Figure 1). We encourage you to make these decisions based on your own knowledge of your students and context, and to view our suggestions as to which year levels each sequence is appropriate for as a rough guide only. Likewise, we have put forward a tentative order in which the sequences might be taught within each domain (again, see Figure 1); however, there is certainly scope to modify this order based on what you deem most beneficial for your students.

Figure 1 Suggested order of the sequences

	ORDER OF SEQUENCES	F	1	2
Number	Counting principles (0–20)	X		
	Structure of number	X	X	X
	Making things equal	X	X	X
	Counting patterns		X	X
	Place value		X	X
Measurement	Informal length measuring	X	X	X
	Time	X	X	X
	Informal approaches to perimeter and area		X	X
	Volume		X	X
Geometry	Recognising polygons	X	X	X
	Objects	X	X	X
	Reasoning with polygons		X	X
	Location and transformation			X
Probability and statistics	Probability and statistics	X	X	X

Although these sequences represent distinct macro mathematical concepts, they are also highly interconnected. The matrix table of the sequences (see page 179) demonstrates how these important concepts and skills are covered by multiple sequences. For example, five of the sequences expose students to *early multiplicative thinking*, whilst all sequences require students to *construct and interpret representations* and to *explain and justify their reasoning*.

OXFORD UNIVERSITY PRESS

Unpacking a learning sequence

We strongly recommend that, before looking at the tasks within a sequence of learning, you carefully read the front section of the sequence. Each introductory section provides an overview of the sequence, summarises several learning suggestions and the associated mathematical focus statements, and details the rationale for the sequence. Within the rationale for each sequence we explain why we think the sequence provides an important learning experience for students, highlight the important mathematical idea(s) we want students to be making sense of, and outline common misconceptions associated with this idea(s). Each introductory section also describes the necessary language for students to develop, provides some suggestions for how the tasks in the sequence might be launched, and includes both a series of specific statements for supporting assessment and a suggested task that teachers might want to use as a pre- and post-assessment task.

Language

The language includes a list of key terminology that students may encounter during the learning sequence, which is important for making sense of the mathematics. We encourage you to introduce the more technical, precise mathematical terminology whenever possible, and to encourage students to use this language, in order to build their mathematical register. This provides students with the language to label important mathematical ideas, and empowers them to be 'thinking, talking and acting' like a mathematician. You may wish to develop an associated 'word wall' alongside the sequence, where new words and important phrases can be added as they are encountered by the students, and which can be referred back to as needed.

Launching tasks

Each sequence contains specific ideas for how tasks in the suggestion might be launched that take into consideration the nature of the mathematical learning involved. However, there are also some more general considerations for launching a task that we outline here.

1   The launch is mainly about posing the task. The challenge for teachers is to pose the task but avoid giving clues about how to solve it or telling the students anything that they can work out for themselves. The rationale for teachers 'holding back from telling' is outlined in the introduction to this book, where we provide an overview of our instructional model and student-centred structured inquiry in more detail (see page v).

2   The launch includes posing and clarifying the language and representation (materials, drawings, symbols) of the task. This also generally involves reading the task to the class or inviting a student to read the task. On occasions, you may wish to emphasise what a *quick maths drawing* looks like, where the goal of the representation is to communicate mathematical information, rather than be artistically creative (e.g. using dots to represent a person or animal).

3   Make your expectations of the way students will work very clear and explicit. For example, *You will be working for 5 minutes independently. After that, I will offer you an enabling prompt if you are stuck. After 10 minutes, you will have a chance to share your thinking with others and collaborate.* The goal is to reduce uncertainty over how the lesson will unfold so students can focus on the mathematics.

4   During the launch, encourage students to imagine their solution before they start the task. Providing opportunities to visualise is a powerful strategy for activating cognition and supporting the development of more abstract mathematical thinking.

5 You may wish to create a story to which the students can relate, as a way of posing the task. For the most part, we have chosen to present tasks where the narrative is not well developed. This is deliberate. We believe that teachers are in the best position to contextualise these tasks in a manner that resonates with their students and context.

6 You may use a preliminary activity to prepare students to think about the mathematical concepts and rehearse fluency skills relevant to the lesson. This type of preliminary activity prepares students for the lesson, and the lesson helps to simultaneously reinforce their fluency and conceptual understanding.

Assessment

There are at least two distinct approaches for assessing students.

1 Choose a task from the sequence, relevant to the level of your students, that you ask the students to do before the beginning of the sequence, and then repeat it one week after the sequence. With a colleague, compare the understanding of the ideas underpinning the task and strategies used as reflected in student work samples.

2 Alternatively, as a team, you might consider foregoing a pre-assessment, and instead focus on formatively assessing students in relation to their understanding of the mathematics as they work through the sequence, using the specific statements to inform assessment as a guide. You may even wish to create a checklist that includes these statements to support your observations of students. Then you might consider choosing one task from the sequence as a follow-up summative assessment, which is administered to students after completing the sequence. You may wish to create a rubric to assess student performance on that task.

In the assessment section for each sequence, we have highlighted possible tasks you might wish to use for this pre- and post-assessment. However, clearly the most appropriate task will depend on your students and how much of the sequence you explored (or plan to explore). In addition to specific statements to inform assessment that are included in each sequence, you should also consider assessing students in relation to important general skills they are developing, specifically how they:

- explain their thinking
- record their thinking and solutions
- articulate their thinking and connection making
- demonstrate persistence on tasks, risk-taking and engagement.

Unpacking a suggestion

A sequence of learning is made up of a series of ordered learning suggestions that serve to organise the tasks in the sequence. Although the suggestions are organised sequentially, you may choose to only do some of the learning suggestions within a sequence. This decision will depend on the year level you are teaching and where your students are at in their mathematical learning, including what might have been covered when you are not teaching with these sequences. We have also seen teachers successfully choose to do part of a sequence early in the year (e.g. do the first three suggestions in Term 1), and then revisit the sequence later in the year (e.g. do the last three suggestions in Term 4).

Each of these learning suggestions begins by outlining an associated mathematical focus statement that serves to anchor the suggestion. This mathematical focus statement should be thought of as pertaining to the *whole* suggestion, rather than just individual tasks. After that, the pedagogical focus for the suggestion is discussed, and a core task (presented in a box) is put forward. Some of the suggestions begin with a proposed preliminary experience 'prior to launching the suggestion'. The goals of this experience are: to introduce important mathematical representations that students will need when working on the tasks in the suggestion; to connect with important prior learning; and to support cognitive activation.

The suggestion concludes by putting forward a series of follow-up tasks under the heading 'Consolidating the learning'. We would generally view a suggestion as being anywhere between two and six lessons in duration, depending on how many consolidating experiences the teacher deems necessary or desirable.

The core task

Usually, the core task is the first task in the suggestion. For core tasks in the number domain sequences, with the exception of the 'Counting principles' sequence, exemplar enabling and extending prompts are included. For most tasks in the other non-number sequences, enabling and extending prompts are not viewed as necessary, as the tasks have been designed to be low floor/high ceiling; that is, both highly accessible to all students and offering opportunities for extension.

Immediately following the core task, we briefly discuss various student approaches to the task and, in particular, anticipated solutions. Combined with the presentation of exemplar enabling and extending prompts (in the number sequences), this anticipation of the mathematical work is intended to support teachers to consider what is involved in forestalling student thinking when working on challenging tasks.

Consolidating the learning

It is recommended that teachers incorporate as many of the follow-up tasks as necessary, either until students obtain a level of mastery in relation to the mathematical focus statement, or until the teacher concludes that students have made as much progress as possible towards the focus statement at this point in their learning. We encourage you to modify these follow-up experiences to make them more or less challenging, based on your observations of student progress on the core task.

You will note that these follow-up tasks generally do not contain suggested enabling and extending prompts. It is our expectation that teachers will be in a strong position to develop such prompts themselves, following our suggestions of exemplar prompts around the core task. Some useful tips for developing enabling prompts include:
- changing the form of representation
- simplifying the language
- changing the size of numbers
- decreasing the number of steps
- helping students connect to past learning
- prompting students to try a particular strategy.
  Similarly, some useful tips for developing extending prompts include:
- changing the form of representation
- increasing the complexity of numbers
- increasing the number of steps

- varying the language
- presenting a similar problem in reverse
- inviting students to make their own version of the task
- finding all solutions
- solving using different strategies
- transferring thinking to another situation (find a generalisation).

In addition to developing prompts, remember the importance of spending some of your planning time anticipating student thinking on these follow-up tasks.

Oxford OWL

Online materials

All BLMs referred to in this book are available online at www.oxfordowl.com.au.

Final thoughts

We believe this book can be used in multiple ways. You may choose to develop your entire mathematics program around this text, using the sequences for the substantial majority of your mathematics instruction. Indeed, this is how many teachers used the sequences in our associated research project. Alternatively, you may decide to use these sequences to complement your existing mathematics program, and perhaps plan to teach one sequence per term. Obviously, this is up to you and your school to decide. We can say with confidence that we have seen both approaches work effectively.

Finally, after looking through the sequences, you might initially feel that your students are not ready to engage with these tasks as currently presented. If students are in their first year of school, we recommend reading Appendix 3, 'Early learning of number' (see page 173), which contains some tasks and activities that can be viewed as preliminary experiences to support engagement with the 'Counting principles' sequence (see page 1). More generally, we would encourage all teachers, particularly those who remain unsure of where to begin with these sequences, to read Appendix 2, 'Overcoming obstacles', where we answer the question: *What are some of the obstacles to teaching with sequences of challenging tasks in the first years of schooling and how might teachers overcome them?* (see page 167).

OXFORD UNIVERSITY PRESS

CHAPTER 1

COUNTING PRINCIPLES (0–20)

OVERVIEW

This sequence is suitable for students in Foundation. The first four suggestions can be done early in the year, whereas the fifth suggestion can wait until a little later. Foundation teachers are encouraged to revisit the suggestions throughout the year with their students. Suggestions in this sequence can also be adapted for students in Years 1 and 2.

This sequence lays the foundation for counting systematically and encourages students to think flexibly about numbers rather than rely on rote counting.

The following is a summary of the suggestions in this sequence.

	MATHEMATICAL FOCUS
The number sequence	There is a stable order of numbers in a count.
Once and only once	Items in a collection are counted once and only once.
One more, one less	Knowing the number before or after a given number can assist in finding the new total of a collection.
Building connections between representations	A quantity can be represented in different ways, including number words, symbols and models.
10 and some more	An efficient way to determine the total number of objects in a collection is to recognise 10 as a set and 'count on' the rest.

RATIONALE

This sequence is based on the belief that the first step in learning to count is being able to say the sequence of numbers, initially counting forwards but moving to counting backwards as well. There are many descriptions of the steps in counting collections. It is commonly agreed that the key counting ideas to be learned are:

- stable order of numbers in a count
- one-to-one correspondence (count each object once and only once)
- conservation of number (how many objects there are does not depend on how the objects are laid out)
- cardinality (the last number counted tells us the number of objects in the collection).

Suggestions in this sequence are intended to be complementary to what teachers do already. Important informal, play-based experiences can be:

- saying and acting out rhymes and reading stories that focus on number (e.g. *The Very Hungry Caterpillar* by Eric Carle)
- playing games that focus on quantitative arrangements (e.g. dominoes)
- having conversations involving comparisons (e.g. *Who has more grapes?*).

In more formal settings, teachers can plan purposeful experiences, such as:

- construction or threading activities using number as a describing word (e.g. make a three-two pattern)
- comparison tasks involving estimation (e.g. comparing two sets, such as a set of 5 objects and a set of 9 objects, and estimating which is larger)
- comparison tasks that require rearranging objects (e.g. *Show me which group has more blocks and how you know this. Move some blocks so that each group has the same number.*)
- finding a given amount (e.g. *Find me four of something.*).

Suggestions in this sequence can be explored multiple times to help students develop flexibility in their thinking about counting systematically.

LANGUAGE

number names to 20, more, less, count, same, between, increase and decrease

LAUNCHING THE TASKS

You might lead discussions intended to connect key learning ideas with student experiences. You could ask questions such as: *Tell me about a time you counted something. Have you seen anyone count in a different way to you? Where do you often see numbers as symbols?*

Suggestions in this sequence can later be used prior to launching other number sequences, such as the 'Making things equal' sequence.

ASSESSMENT

The following task is suggested for pre- and post-assessment: Suggestion 5: 'Three Little Pigs'.

The following are some specific statements to inform assessment.
Students are learning about counting when they can:

- identify and verbalise numbers in a sequence (forwards or backwards)
- identify and compare numbers on dice without counting
- count systematically and explain their thinking
- identify and compare numbers with different representations and explain their thinking
- solve problems related to counting systematically.

SUGGESTION 1: THE NUMBER SEQUENCE

MATHEMATICAL FOCUS

There is a stable order of numbers in a count.

PEDAGOGICAL FOCUS

These following experiences are intended as practice with counting forwards and backwards. The first two are about the sequence of the number names, while the third and fourth suggestions are more about conceptual subitising (immediate recognition) of the patterns on the dice. Sight recognition of quantities up to five or six is an important stage for students, as it not only supports more sophisticated counting and enumeration but also accelerates the development of addition and subtraction. There is no need for students to do this quickly at first; the intention is that they speed up over time.

Once students are ready for more challenge (including those in Years 1 and 2), there are variations to the first suggestion, such as starting from a number other than 1, skip-counting, using larger numbers, money and fractions.

You may like to start this task as a class, before students work in groups.

DIRECTIONS

In small groups, students stand in a circle and count from 1.

Each student uses their hand to indicate the direction of who counts next (to the left or to the right).

This can be used for counting forwards or backwards.

Comment

It is suggested that the count continues forward from 1 despite the direction changing.

CONSOLIDATING THE LEARNING

There are three further tasks that explore counting in a sequence or number patterns. The first consolidates the learning from the first task, and the other two tasks extend that learning, moving to conceptual subitising.

CLAP COUNTING

The teacher claps while saying the first few numbers in the sequence, then stops saying the numbers aloud and keeps clapping. The students count in their heads and say the finishing number together.

This can be forwards from 0 or back towards 0. Larger numbers may be used to extend student thinking.

TENZI

Each of 4 players have 10 dice. The goal is to continue rolling the dice until all dice show the same number.

Each player should choose a number and then roll the dice. If their number comes up, they keep those dice aside. Players keep rolling the other dice, putting the dice with their number aside on each roll. The winner is the first to have all their dice showing the same number.

ALTERNATIVE WAYS OF PLAYING TENZI

Each player rolls one dice to allocate a number (from 1 to 6). Explain that this is their number and then turn the dice (physically) until all dice show the allocated number (e.g. 5, 5, 5, 5, 5, 5 …).

Another suggestion could include rolling or physically turning the dice and arranging them in order (1 to 6).

SUGGESTION 2: ONCE AND ONLY ONCE

MATHEMATICAL FOCUS

Items in a collection are counted once and only once.

PEDAGOGICAL FOCUS

Once students have heard the number names, the next step is to focus on counting once and only once, especially counting systematically and being able to explain the system they use.

For the first two tasks, teachers could use the prompt *What question are you asking?* rather than posing the questions themselves. The idea of counting without touching is to emphasise the need for a system. It also simulates many real counting situations in which it is not possible to touch the objects we are counting.

Without a system for counting, students are more likely to double count, continue counting when objects are in a circle, or omit objects. A common counting error that occurs when students rely on rote counting is saying the number names faster (or slower) than the pace at which they refer to the associated objects. The suggestions in this sequence intend to develop the concept of one-to-one correspondence, which is an important prerequisite for relational counting (where students give correct number names as objects are counted in sequence and can describe the total number of objects irrespective of position or characteristics).

The fourth and fifth tasks consolidate the idea of one-to-one correspondence but also explore conservation of number (how many objects there are does not depend on how the objects are laid out). Encourage students to be systematic with their counting and justify their thinking.

Prior to launching the first task, you could tune students in by playing a brief game of 'Directions'. Initially, teachers could use collections of counters to launch the task.

COUNTING WITHOUT TOUCHING 1

What question are you asking?

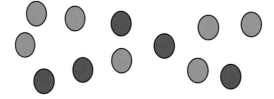

Comment

Students may use a variety of strategies to count, including one colour first, count from left to right, groups, lines or following a shape (e.g. fish).

OXFORD UNIVERSITY PRESS

Enabling prompt

How many blue dots are there?

Extending prompt

Find two fast ways to compare sets (such as blue and green).

CONSOLIDATING THE LEARNING

There are five further tasks that explore the idea of counting once and once only. Encourage students to be systematic in their counting. The final suggestion is perhaps the most important and can be done many times.

COUNTING WITHOUT TOUCHING 2

What question are you asking?

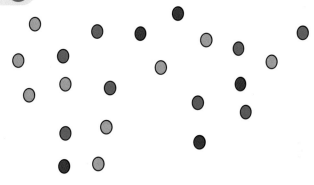

COUNTING IN A CIRCLE

How many dots do you see? How can you be sure?

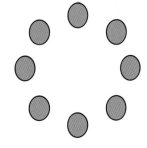

COUNTING IN A CIRCLE AND LINE

Guess which has more dots, the circle or line? Why do you think this?

Count how many dots make up the circle and line.

Which has more dots? How can you be sure?

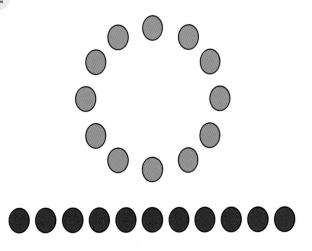

COUNTING IN A CIRCLE, RECTANGLE AND LINE

Guess which has more dots, the circle, rectangle or line? Why do you think this?

Count how many dots make up the circle, rectangle and line.

Which has more dots? How can you be sure?

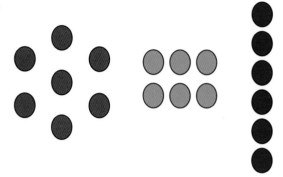

COUNTING TOGETHER

Create a collection of objects.

Your partner points to your objects without touching each of them and without counting them.

Then your partner counts your objects without touching them.

Then you both count them together (with touching).

SUGGESTION 3: ONE MORE, ONE LESS

MATHEMATICAL FOCUS

Knowing the number before or after a given number can assist in finding the new total of a collection.

PEDAGOGICAL FOCUS

This suggestion is intended to prompt students to realise that it is not necessary to count the whole collection again once something changes, but to count from the previous total instead.

The first task introduces the principle of dynamic counting to develop flexibility with counting. The term 'dynamic' captures the idea that the total number of objects in the set is continuously changing.

For the first task, imagine that four students are sitting in a group, with one of them (or an adult) acting as the 'dealer'. The dealer progressively adds or removes one counter from the table, and the other group members must say the total together. The pace of the activity illustrates to students that they should not recount the total number of counters. Instead, students use their knowledge of the previous total and the action of the dealer (adding or removing 1) to determine the new total. It is important for the dealer to go beyond the numbers that the students can recognise immediately, to numbers for which some strategy is needed (such as 7, 8, 9). The students in the group can take turns being the dealer. The emphasis is on getting students to think about how numbers in a sequence relate to each other, rather than just rote counting. This can be done many times, sometimes in small groups, sometimes as a class.

There are no enabling and extending prompts suggested. In fact, even if students are not ready for the experience, they will benefit from seeing and hearing the other students counting.

The emphasis is on getting students to think about how numbers in a sequence relate to each other, rather than just rote counting. Variations of the 'Dynamic counting' task for students in Years 1 and 2 could include using money instead of counters.

When launching the first task you may like to start as a class before students work in groups.

DYNAMIC COUNTING

Counters are presented one at a time, sometimes increasing, sometimes decreasing, and students count together.

CONSOLIDATING THE LEARNING

There are three further tasks that explore ideas related to dynamic counting. Two tasks consolidate the learning from the first task. The final task, 'Reading numbers together', is harder and will take longer for students to develop fluency.

ONE-LESS-THAN DOMINOES

Instead of matching ends, a new domino can only be matched if it has one less than the pattern shown.
This game can be played for 1 more, and 2 more or less than.

ONE DIFFERENT SNAP

A variation on the game of snap. Instead of matching cards, the players snap when the cards are different by 1.

READING NUMBERS TOGETHER

The class says the numbers together (starting at either the top or bottom), or they can add 1, subtract 1, etc. You may like to increase the size of the numbers when students are ready for a further challenge.

3

5

6

2

9

8

1

7

4

MATHEMATICAL FOCUS

A quantity can be represented in different ways, including number words, symbols and models.

PEDAGOGICAL FOCUS

Students meet numbers first as spoken words that are connected to a representation (2 people, 3 dogs). Later the intention is that they also connect the symbol to the representation and the word.

BLM 1: Making different representations presents four representations of the numbers from 5 to 8: the numeral, a pattern, the dots in a circle and an additive representation. The worksheet should be cut to make individual cards for sorting. Of course, the intention is that you can add other representations, such as the words or even a number sentence (3 + 2), and you can use this for other numbers. You can also provide a 4 by 4 array onto which students can place the cards.

No prompts are suggested with this task since students are manipulating the cards, although smaller and larger numbers can be used.

Prior to the launch, you could briefly revisit the subitising activities from Suggestion 1, such as 'Tenzi'.

MATCHING DIFFERENT REPRESENTATIONS

Sort the cards into groups.
Describe each of your groups

CONSOLIDATING THE LEARNING

There are six further tasks exploring the connection between spoken numbers and their representations.

FOR 9 AND 10

Create the cards that would be needed for the numbers 9 and 10.

WHAT NUMERAL GOES WITH THESE CARDS?

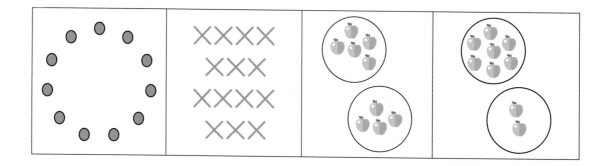

MEMORY

Using cards from the first task, play memory for matching pairs. More cards could be created for this game.

FIND YOUR PARTNER

Everyone is given one card and they walk around (without talking) to find their partner(s).

GO FISH

Deal out 5 cards and have the students make pairs, etc.

ORDERING DIFFERENT REPRESENTATIONS

Arrange these cards in order, based on the number they represent.

7	eight	(gift boxes)	(apples)	XXX XXXX XXX
(smiley faces)	1	two	(octopuses)	five

SUGGESTION 5: 10 AND SOME MORE

MATHEMATICAL FOCUS

An efficient way to determine the total number of objects in a collection is to recognise 10 as a set and 'count on' the rest.

PEDAGOGICAL FOCUS

When starting school, many students can count to 20. Many other students are ready to extend their counting range. The tasks in this suggestion are intended to prompt student thinking beyond 10. The goal is for students to see that they do not have to count the '10'. Composing and decomposing numbers from 11–19 lays the foundations for place value concepts.

This suggestion may be revisited throughout the year to consolidate student learning. Similarly, changing the context will provide further learning experiences, for example, 'Pencil (2)'.

Prior to launching the 'Three Little Pigs' task, students may benefit from an initial experience, building the idea that they can count without counting the '10'. The following two experiences could be done as a whole class or individually. You could ask students how they might count the dots systematically and compare approaches.

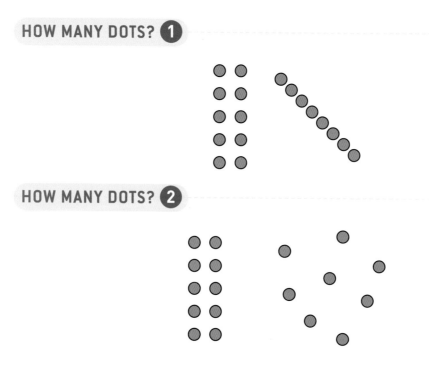

Comment

Encourage students to imagine their solution before they start each task. The 'Three Little Pigs' task can be done as a class activity, with students being invited to show how many sticks one or other pigs might have, including counting them.

THE THREE LITTLE PIGS

The Three Little Pigs have gathered sticks.

Each of the Little Pigs has more than 10 sticks but less than 20 sticks.

They want to work out how many sticks each of them has.

How can they count their sticks quickly?

Draw the sticks so they can be counted quickly.

Comment

Students may use a variety of approaches to count their sticks, starting from grouping with 10 and the remaining sticks organised in such a way to support subitising (for example, 10 and 6 more could make 16).

OXFORD UNIVERSITY PRESS

CONSOLIDATING THE LEARNING

There are three further tasks consolidating the idea of counting without needing to count the '10'. Teachers are encouraged to revisit these tasks and adapt as appropriate by providing students with opportunities to compose and decompose teen numbers.

THE BRAINIE BOX

I have 3 boxes of Brainies. Each box has a different number of Brainies.

In each Brainie box there are between 15 and 20 Brainies.

How many Brainies might be in each of my boxes?

Draw the Brainies so you count them quickly.

Students may find it helpful to record their ideas on a sheet.

Brainie box 1	Brainie box 2	Brainie box 3

PENCILS 1

Three groups of students share a container of pencils to make a poster.

Each container has more than 10 pencils, but less than 20.

How many pencils might there be in each container?

Draw the pencils in their containers so you can count them quickly.

Give as many possibilities as you can.

PENCILS 2

45 pencils in a container were shared between three groups of students.

Each group had at least 10 pencils.

How many pencils might each group have?

Draw the pencils in the three groups so you can count them quickly.

Give as many possibilities as you can.

CHAPTER 2

STRUCTURE OF NUMBER

OVERVIEW

This sequence is suitable for students from Foundation to Year 2. The first four suggestions are suitable for all years. The goal is to encourage students to develop an understanding that goes beyond counting and focuses on how numbers are structured (e.g. exploring 'ten-ness'). Students will be provided with experiences in which they consider conceptual subitising and part–whole relationships, and link these ideas to an understanding of addition and subtraction. The emphasis is on students developing flexibility with numbers.

The following is a summary of the suggestions in this sequence.

	MATHEMATICAL FOCUS
Conceptual subitising (part–whole thinking)	The number of objects in a collection can be recognised without counting.
Introducing part–whole thinking (2 parts)	The number of objects in a collection can be partitioned into two different parts.
Extending the part–whole idea (3+ parts)	The number of objects in a collection can be partitioned into three or more different parts.
Introducing addition and subtraction	Numbers can be added and subtracted.
Connecting addition and subtraction	Addition and subtraction are connected.
Extending addition and subtraction	Recognition of patterns within number facts can assist to add or subtract more than two numbers efficiently.

RATIONALE

This sequence is intended to move students beyond counting by ones and develop an appreciation for the structure of number. For example, in order to find the number of objects in a group, students need to recognise that numbers are constructed of smaller parts. This involves partitioning and combining numbers.

Initially, students explore these ideas using concrete materials and pictorial representations, before moving to formally recording their thinking, and then to representing this thinking using number sentences.

The first three suggestions can be explored multiple times in the first year of schooling to help students develop flexibility with number. By Years 1 and 2, this flexibility should correspond to working with number sentences involving the addition and subtraction operations. It is recommended that teachers take time to explore/revisit the first three suggestions to ensure a strong foundation for this more formal mathematical work. Teachers need to consider modifying the numbers in the problems to ensure they are suitably challenging for their particular cohort of students.

OXFORD UNIVERSITY PRESS

One powerful tool in Suggestion 5 that supports students' exploration of the relationship between addition and subtraction, and mental computation more generally, is the open-number line. We recommend introducing this tool alongside other representations, such as part–whole boxes, to help deepen students' understanding of the structure of number.

Suggestions in this sequence can be revisited multiple times to help students develop flexibility in their thinking about number, and deepen their understanding of the relationship between addition and subtraction. Thus, it is important to explore these concepts together.

LANGUAGE

subitise, partition, combine, separate, add, subtract, difference, number sentence, equation, represent, record

ASSESSMENT

Suggested pre- and post-assessment task: Suggestion 2: '11 pencils'. Use different numbers in the post-assessment task.

The following are some specific statements to inform assessment.
Students are learning about structure of number when they can:
- demonstrate that arranging objects into known patterns can support *conceptual subitising* (for a given number, e.g. 8), present examples of arrangements that support subitising, and examples of arrangements that do not support subitising
- explain and justify that a whole can be divided into many different parts
- record multiple solutions to part–whole problems
- describe and record mathematical situations involving addition and subtraction using number sentences
- use the part–whole model to explain how addition and subtraction are related
- partition numbers to support them to solve addition and subtraction problems efficiently
- work systematically and identify patterns to make sure all solutions have been found.

SUGGESTION 1: CONCEPTUAL SUBITISING (PART–WHOLE THINKING)

MATHEMATICAL FOCUS

The number of objects in a collection can be recognised without counting.

PEDAGOGICAL FOCUS

For the tasks that require students to subitise, show the images quickly so that students cannot count by ones. Students should be able to recognise collections up to 4 initially without the need for counting (perceptual subitising). Through knowledge of patterns (e.g. dice) this can be quickly extended to conceptual subitising. The idea is that students hold the images mentally and manipulate the groups of objects. The 'Summer time is fly time!' and 'Delicious doughnuts' tasks enable students to discover and explore these important principles. Conceptual subitising supports part–whole/additive thinking and can be extended into multiplicative thinking.

Prior to launching the suggestion, introduce the term 'subitise' (to know how many straight away) and provide students with a sense of what it means to subitise. This can be done by presenting a few simple dot patterns on cards (e.g. 3 dots) and asking students, *How many dots do you see?* and then, once they respond, *How did you know?*

Students may arrange their flies in a line – an important representation to support accurate counting, but an automatic response that can make subitising difficult. It is important to exercise patience and restraint, particularly with this first task, to make it clear to students that you expect them to figure it out themselves.

Providing story contexts for the tasks is useful, such as flies in summer, or reading a picture storybook such as *Tiny Little Fly* (Michael Rosen).

SUMMER TIME IS FLY TIME!

I saw 10 flies on my bedroom wall, and I knew how many there were straight away without counting them.

Can you draw what the 10 flies on the wall might have looked like? Now draw them a different way. Which picture do you think makes it easier to know there were 10 flies on my wall?

Comment

There are many possible solutions to this task. The acceptability of a given solution depends on how the student defines 'straight away'. We are looking for solutions where the whole 'jumps out' from the parts (e.g. 5 and 5 using the dice pattern; 6 and 4; a ten-frame pattern). The purpose of getting students to draw two pictures is to allow them to make a comparative assessment about which picture makes the whole '10 flies' more obvious.

Enabling prompt

What if there were 6 flies on my wall?

Extending prompt

What if there were 16 flies on my wall?

CONSOLIDATING THE LEARNING

There are three additional tasks presented in this suggestion that build on from and consolidate the learning from the first task. Teachers should modify the numbers, so they are appropriately challenging for their students.

FISH TANK

I have 11 fish in my fish tank at home.

Draw three different pictures of my fish in their fish tank. Which picture do you think makes it easiest to know how many fish I have? Explain your reasoning.

DELICIOUS DOUGHNUTS

I saw two doughnuts. One had 19 sprinkles on it and the other had 7. I knew the first doughnut had 19 sprinkles almost straight away. But I had to count the doughnut with 7 sprinkles. Draw what the doughnuts might have looked like.

BIRTHDAY CAKE

I went to my cousin's birthday party. She wouldn't tell me how old she was turning, but I could tell straight away by looking at the pattern the candles made on her cake. How old might my cousin have been turning? What might the pattern of candles on the cake have looked like?

SUGGESTION 2: INTRODUCING PART–WHOLE THINKING

MATHEMATICAL FOCUS

The number of objects in a collection can be partitioned into two different parts.

PEDAGOGICAL FOCUS

BLM 2

Oxford OWL

The focus is on different ways to partition a number (recognising that a number is constructed of smaller parts). Students may use manipulatives to split the number (e.g. Unifix). The tasks have multiple answers. You may wish to provide a template for students – see **BLM 2: The cubby house**. Encourage students to imagine a solution before inviting them to use other ways, such as manipulatives/drawings.

Prior to launching the task teachers could read a picture storybook (e.g. *When I Was Little, Like You* by Mary Malbunka). To launch the suggestion, teachers might like to create their own story.

As students work on their solutions, take note of their recording. For example, whether they are recording in systematic ways recognising patterns, or in an ad hoc manner. Questions to prompt students while they explore the problem include, *How do you know you have found all the possible solutions? Can you convince us that you have found them all? What do you notice about how the solutions are recorded?*

CUBBY HOUSE

9 friends are playing in a cubby house.

Some of the friends are playing inside the cubby house and some are playing under the cubby house.

Draw a picture to show how many friends might be inside and how many friends might be under the cubby house.

Give as many answers as you can.

Comment

There are 8 possible solutions to this task (1 inside, 8 under ... 8 under, 1 inside). Some students might interpret 'some' to mean more than 1 (which would limit the problem to six solutions). This ambiguity is okay. It can lead to the question: *What would it mean to play on your own in this scenario?*

Enabling prompt

What if there were six friends inside the cubby house? How many friends would be under the cubby house?

Act out different situations.

Extending prompt

There are eight possible pictures. Can you draw them all?

CONSOLIDATING THE LEARNING

Two further tasks, involving slightly larger numbers, consolidate the learning of the first task.

11 PENCILS

Together, Eden and Odin have 11 pencils. How many pencils might they each have? Give as many different answers as you can.

MORE RED THAN GREEN

I have 15 counters. I have more red than green counters. Draw what this might look like.

SUGGESTION 3: EXTENDING THE PART–WHOLE IDEA

MATHEMATICAL FOCUS

The number of objects in a collection can be partitioned into three or more different parts.

PEDAGOGICAL FOCUS

The focus is on different ways to partition a number. Moving from part–whole problems with two parts, to part–whole problems with three or more parts substantially increases the number of solutions. For other problems in the suggestion, it is predominantly about being systematic (e.g. 'Cubby house again').

Prior to launching the task, students could sing a relevant song (e.g. 'Old MacDonald') or look at pictures of farm animals. As part of the launch, ask students to imagine what animals might be in the photo, before drawing them.

OXFORD UNIVERSITY PRESS

In a photo of a farmyard, you can see 12 legs.
Draw what the animals might be.

Comment

It is expected that students will provide answers incorporating 2-legged and 4-legged creatures (e.g. 6 chickens, 3 cows, 2 cows and 2 chickens, 1 cow and 4 chickens); however, there is room for creativity here. For example, a student might reason that it could be an ant farm, an octopus farm or a zoo. Solutions can be achieved through a combination of creativity (e.g. a three-legged dog) and being systematic (e.g. cows and chickens).

Enabling prompt

I have two pets. They have 6 legs altogether. What pets might I have?

Extending prompt

How many different combinations of animals are possible?
How do you know?

CONSOLIDATING THE LEARNING

Three further tasks consolidate the idea that numbers can be partitioned in different ways. It is important that students are encouraged to find a range of possible solutions.

THREE CHILDREN

In my family there are three children. The total of their ages is 13 years. All their ages are different. What might their ages be? (Give more than one possible answer.)

CUBBY HOUSE AGAIN

Thirteen friends are playing in a cubby house.
Some of the friends play inside the cubby house, some play under the cubby house and some play outside the cubby house.
Draw a picture to show how many friends might be inside, how many might be outside and how many might be under the cubby house.

Comment

Encourage students to find as many possibilities as they can. There are 66 possibilities altogether. For example, there could be 11 friends inside the cubby house, 1 friend under and 1 friend outside; there could be 1 friend inside, 11 friends under and 1 friend outside; and 1 friend inside, 1 friend under and 11 friends outside, and so on.

Five students have 21 counters between them. Each student has a different number of counters. How many counters might the students each have?

SUGGESTION 4: INTRODUCING ADDITION AND SUBTRACTION

MATHEMATICAL FOCUS

Numbers can be added and subtracted.

PEDAGOGICAL FOCUS

The focus of this suggestion is for students to become fluent with known number facts and to encourage them to be more systematic with their thinking when working with addition and subtraction.

It is expected that students will (begin to) represent their thinking using symbols and record their solutions using number sentences.

Prior to launching the task, teachers could use work samples from a previous suggestion and, with input from students, record some of the pictorial representations (e.g. 'Cubby house' task) as number sentences (e.g. $1 + 8 = 9$; $2 + 7 = 9$; or $1 + 1 + 11 = 13$, $1 + 2 + 10 = 13$ for the 'Cubby house again' task). This would demonstrate to students how number sentences are powerful ways of recording our mathematical thinking.

VARIATIONS ON TENZI ❶

Roll 10 dice. Find combinations of dice that add up to 10. How many different ways can you make 10 from the dice you rolled? Record your solutions in number sentences.

Comment

Possible solutions using just two dice are 5 and 5; 6 and 4. Students quickly realise that many other possibilities exist once three or more dice are used. An interesting question (and one which we have not solved ourselves) is what values on 10 dice would enable the greatest number of unique possibilities for making 10. For example, if a student rolled 10 ones, there is only one way to make 10 ($1 + 1 + 1 \ldots = 10$). If a student rolled 10 sixes, there are no ways of making 10. However, if a student rolled 1, 1, 2, 4, 5, 5, 6, 6, 6, 6, there are four ways of making 10 ($6 + 4$; $5 + 5$; $6 + 2 + 1 + 1$; $5 + 4 + 1$). What about the combination: 1, 1, 1, 2, 2, 2, 3, 4, 5, 6?

Enabling prompt

I rolled 5 and 3. How many more do I need to make 10?

Extending prompt

Keep rolling your dice until they can all be used to make combinations to 10 (so you have no dice left over). Compare your results to a classmate's. Are the combinations you have made the same or different? Why do you think this is the case?

CONSOLIDATING THE LEARNING

The additional tasks extend students' learning and encourage the use of mental computation and strategic thinking.

RACE TO 11

Two players take turns at adding either 1 or 2. Start at 0. The winner is the person who says 11. Note: students can place counters in a ten-frame or draw on a number line.

RACE TO 0

Two players take turns at subtracting either 1 or 2. Start at 11. The winner is the person who says 0.

VARIATIONS ON TENZI 2

Roll 10 dice. Find pairs that add to 7.

VARIATIONS ON TENZI 3

Roll 10 dice. Find pairs with a difference of 2.

SUGGESTION 5: CONNECTING ADDITION AND SUBTRACTION

MATHEMATICAL FOCUS

Addition and subtraction are connected.

PEDAGOGICAL FOCUS

The focus of this suggestion is to explore a different representation of the relationship between addition and subtraction, in a more abstract way. Some tasks have multiple answers while others have single answers. We recommend teachers explore some 'Empty boxes' tasks before the story problems.

The story problems have been developed to avoid students attempting to tackle worded problems in an automated way, such as relying only on 'key words', or the order in which the numbers are presented. There is value in students exploring in depth the relationships between a suite of connected problems.

Prior to launching the worded problems, emphasise the power of drawing a picture or representing the mathematical situation in some other meaningful way (e.g. Unifix, diagrams). These concrete models will support students to engage with both the worded problems and the more abstract empty boxes representation. It is important to emphasise to students that unknown numbers can be represented in many different ways (e.g. empty boxes, in later schooling, letters etc.).

Find the answers to these questions, and show how you worked out your answers using empty boxes or empty number lines:

In a family, there are 5 boys and 8 girls. How many children are there altogether?

In a family, there are 8 girls. If there are 13 children altogether, how many boys are there?

In a family with 13 children, 5 of the children are boys. The rest of the children are girls. How many children are girls?

How are these questions the same and how are they different?

Comment

The key similarity across the questions is that they are describing the same mathematical situation (13 children in a family: 5 boys and 8 girls). However, in each of the questions, a different part of the question is 'unknown'. The first question is addition, whereas the second and third questions can either be described as missing addend problems ($8 + \square = 13$) or subtraction problems ($13 - 8 = ?$). Getting students to connect these two different representations allows them to see how addition and subtraction are fundamentally connected, using the part–whole conceptual model to support their understanding.

Enabling prompt

Choose 13 counters to represent the children. Use one colour to represent the girls, and a different colour to represent the boys.

Extending prompt

Our large family of 13 just got bigger as mum gave birth to a set of quintuplets! Work through these questions again with this new information.

CONSOLIDATING THE LEARNING

The 'Empty boxes' tasks (at least 1 and 2) should precede the 'Chickens' task. The other worded problem tasks consolidate the learning from the 'Chickens' task and the relationship between addition and subtraction in authentic contexts. The 'More than' and 'From here to there' tasks focus on using number line representations to explore similarities and differences.

 EMPTY BOXES **1**

This shows $3 + 4 = 7$. It also shows $7 - 4 = 3$ and $7 - 3 = 4$.

What numbers might go into the empty boxes?

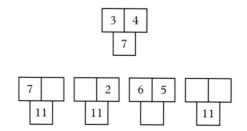

EMPTY BOXES ❷

What numbers might go into the empty boxes?

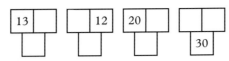

EMPTY BOXES ❸

What numbers might go into the empty boxes?

EMPTY BOXES ❹

What numbers might go into the empty boxes?

CHICKENS

Find the answers to these questions, and show how you worked out your answers using empty boxes or empty number lines.

a   On a farm, there are 31 white chickens and 10 brown chickens. How many chickens are there altogether?

b   On a farm, there are 10 chickens. If I want to have 41 chickens, how many more chickens do I need?

c   On a farm with 41 chickens, 10 of the chickens are white. The rest of the chickens are brown. How many chickens are brown?

   How are these questions the same and how are they different?

TREES

Find the answers to these questions, and show how you worked out your answers using empty boxes or empty number lines.

a   In a plantation, there are 20 eucalypt trees and 37 wattle trees. How many trees are there altogether?

b   In a plantation, there are 57 trees. Some are eucalypts and some are wattles. If I have 20 eucalypt trees, how many of my trees are wattles?

c In a plantation, there are 20 wattle trees and 57 trees altogether. If the rest of the trees are eucalypts, how many eucalypt trees are there?

How are these questions the same and how are they different?

MORE THAN

Find the answers to these questions and show your working on an empty number line.
- What is 20 more than 37?
- 37 is 20 more than what number?
- How much more than 20 is 37?

What is the same and what is different about these questions?

VEGGIE GARDEN

Find the answers to these questions and show your working on an empty number line.

a We are growing tomatoes and capsicums in our vegetable garden. There are 52 plants altogether. If there are 32 tomato plants, how many capsicum plants are there?

b We are growing tomatoes and capsicums in our vegetable garden. If there are 32 tomato plants, and 20 capsicum plants, how many plants are there altogether?

c We are growing tomatoes and capsicums in our vegetable garden. There are 52 plants altogether. If there are 20 tomato plants, how many capsicum plants are there?

What is the same and what is different about these questions?

FROM HERE TO THERE

Show two different solutions for this question. On the number line, write down the jump and number that the jump lands on. Draw a separate number line for each solution.
- Go from 23 to 55 in 4 jumps.

23   55

SUGGESTION 6: EXTENDING ADDITION AND SUBTRACTION

MATHEMATICAL FOCUS

Recognition of patterns within number facts can assist in adding or subtracting more than two numbers efficiently.

PEDAGOGICAL FOCUS

The focus for this suggestion is on making addition and subtraction easier when adding more than two numbers through looking for patterns and changing the order in which the numbers

OXFORD UNIVERSITY PRESS

are added. For instance, recognising the power of ten facts (e.g. 7 + 3; 8 + 2) will assist with the addition of a sequence of numbers. The idea is for students to partition and regroup numbers formally through applying the associative properties of addition.

It is assumed that students are familiar with tens facts and have some appreciation of these patterns. It is also assumed that students are familiar with numbers to 100, and can recognise that 100 + 30 will equal 130. Even if students do not have a full appreciation of the place value concept, it is anticipated that this response can be generated through some understanding in language of how we construct numbers beyond 100 (i.e. we state 100 and 30 to represent 130). Throughout this suggestion, there should be a strong emphasis on mathematical reasoning and on students explaining their strategies.

Prior to launching the worded problem, we recommend the 'Finding shortcuts' task.

When launching the tasks in this sequence, encourage students to read and reflect on the problem in its entirety and look for patterns before trying to solve it. Getting students to think this way can be challenging, even if they are mathematically highly capable.

CHRISTMAS SHOPPING

Jeffrey did some Christmas shopping for his two sisters. He decided to get them both tickets to see Katy Perry in concert. The tickets cost $99 each. He also got his dog, Sook, a plastic bone for $2. How much money did he spend on his Christmas shopping?

Comment

There are two challenging aspects to the 'Christmas shopping' task. First, students need to understand the mathematical situation presented in the story problem and to represent it in some manner (e.g. 99 + 99 + 2). Second, they need to recognise that partitioning the two into two ones and changing the order in which the numbers are added can support their calculation (99 + 1; 99 + 1).

Alternatively, solutions include compensation-style strategies [(100 + 100 − 2) + 2 = 200], which are only marginally less efficient than partitioning strategies. Some students might try using the split strategy 90 + 90 + 9 + 9 + 2, or the addition algorithm. Often these students are mathematically highly capable. The challenge is to encourage these students to consider more efficient approaches.

Enabling prompt

19 + 19 + 2 =

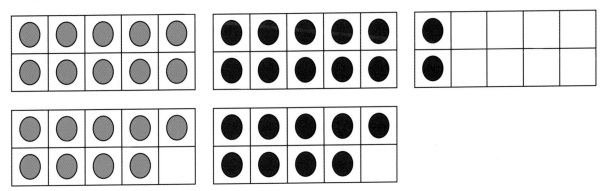

Extending prompt

Jeffrey forgot that his 4 cousins and their dog Fletcher were also going to be at his house on Christmas day. He decided to buy his cousins tickets to Bounce, which cost $49 each. For Fletcher, he bought a new collar. How much more money did Jeffrey have to spend?

CONSOLIDATING THE LEARNING

The 'Finding shortcuts' task below can be used prior to the main task, 'Christmas shopping', whereas the three additional tasks are intended to reinforce the use and power of the tens facts, regrouping numbers using the associative property and place value knowledge.

FINDING SHORTCUTS

Work out the answer to $3 + 5 + 35 + 37$ in your head.

What advice would you give to a friend about how to work out the answers to questions like these in their head?

MORE SHORTCUTS

First, do each question in your head and write the answer, then explain how you worked it out.

$5 + 5 + 15 + 15 =$

$3 + 4 + 16 + 17 =$

$1 + 18 + 9 =$

$2 + 3 + 48 =$

$1 + 1 + 1 + 19 + 19 + 19 =$

$60 - 19 - 19 - 19 - ? = 0$

MORE MISSING NUMBERS

Use a shortcut to work out the missing number in each sentence. Explain your reasoning.

$1 + 2 + 29 + ? = 60$

$100 = 1 + 49 + 2 + ?$

$5 + 25 + ? + 25 = 60$

$? + 9 + ? + 8 = 20$

$? + 28 + ? + 29 = 60$

$60 - 29 - 29 + ? = 0$

CONNECTING JUMPS WITH EQUATIONS

Show two different ways to add 118 by jumping forward from 132 in three jumps. For each of your jumps, write down the number that the jump lands on.

Describe your jumps using equations.

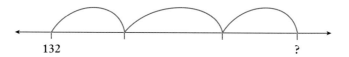

CHAPTER 3

MAKING THINGS EQUAL

OVERVIEW

This sequence is suitable for students from Foundation to Year 2. The first three suggestions are suitable for all years, including Foundation. There may be advantages in students experiencing Suggestions 1, 2, and 3 in the first year of school but at different times of the year. The latter suggestions (4, 5 and 6) are more suited to Years 1 and 2, although those students would benefit from the earlier suggestions as well. Since the tasks have both low floors and high ceilings, there are no enabling and extending prompts provided. All tasks allow students opportunities to explain their thinking.

It is recommended that students experience the 'Counting principles' sequence prior to this one.

The following is a summary of the suggestions in this sequence.

	MATHEMATICAL FOCUS
Different possibilities for making collections the same	Two collections can be compared and then made the same by adding or removing items.
Making collections equal	Collections can be made equal by adding or subtracting.
Recording our thinking using number sentences	Number sentences can be used to record how collections on ten-frames were made equal.
Balancing equations	Both sides of an equation need to be equivalent when the equals sign is used.
Balancing equations beyond 20	There is a relationship between the numbers either side of the equals sign.
Principles of addition and subtraction	Patterns can be used to make sense of addition and subtraction.

RATIONALE

From a young age, children have a good sense of equality based on experiences of 'fairness' and 'sameness'. However, few experiences build the notion of 'the same as' that encourages students to describe the relationship of two quantities as being the same (equal) or different (unequal).

The focus in this sequence is on equivalence and the meaning of the equals sign. It provides a way for students to experience addition and subtraction in a practical setting. It moves students towards formal recording of solutions and to thinking about equivalent number sentences and the relationship between numbers.

The idea of equivalence is best developed concretely using kinaesthetic approaches, tactile objects and visualisations to reinforce the notion of 'balance' of the equals sign. Subsequently, students can then explore numbers and then variables (such as a symbol or letter for the unknown).

LANGUAGE

same as, equals, balance, more than, less than, add, subtract, take away, equation, number sentence

LAUNCHING THE TASKS

The following experiences from the 'Counting principles' sequence could be incorporated in the launch to support student thinking when counting:

- 'Dynamic counting' and 'Reading numbers together' as a way of simplifying counting after a small change, and fluency with counting sequences.
- 'Counting without touching' helps with one-to-one matching as well as 'Clap counting' and 'Directions' for counting systematically.
- 'Tenzi' as a way of subitising conceptually on the ten-frame.

  It is important to keep preliminary learning experiences brief.

ASSESSMENT

Suggested pre- and post-assessment task: Suggestion 1: 'Cakes (3)'.

The following are some specific statements to inform assessment.
Students are learning about equivalence when they can:

- compare two collections and show how they can be made equal
- model, represent and describe how to make collections equal
- create, explain and systematically record several possibilities for equivalent equations, including using symbols
- transfer learning from one experience to another.

SUGGESTION 1: DIFFERENT POSSIBILITIES FOR MAKING COLLECTIONS THE SAME

MATHEMATICAL FOCUS

Two collections can be compared and then made equal by adding or removing items.

PEDAGOGICAL FOCUS

The tasks have multiple possible answers that the students can create using physical models if needed. Students should have opportunities to make decisions about an experience they have had (or can see). Some might count items individually and compare, while others can explore and exhaust possibilities.

CAKES 1

Make the number of cakes on each plate the same (without removing any cakes from the left-hand plate).

Give as many possibilities as you can (each time starting from the set-up shown in the picture).

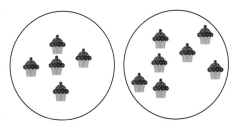

Comment

A variety of solutions can be found without removing cakes from the left plate. There could be 2 cakes removed from the right plate, leaving 5 cakes on each plate, or 1 cake removed from the right plate and placed on the left plate to have 6 cakes on each plate. Some students might seek to create new cakes to add to the plates, such as 2 cakes added to the left plate to have 7 cakes on each plate. Affirm all correct solutions.

CONSOLIDATING THE LEARNING

There are three further tasks that can be used to consolidate the learning. They are essentially the same task using different numbers. It is expected that most students can find multiple solutions. 'Cakes (4)' is more complex than the others.

For each of the consolidating tasks, students are expected to give as many possibilities as they can without removing any cakes from the left-hand plate.

CAKES 2

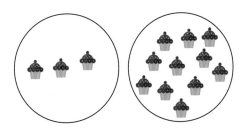

CAKES 3

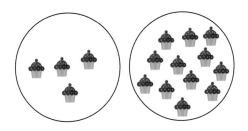

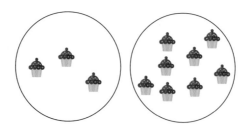

SUGGESTION 2: MAKING COLLECTIONS EQUAL

MATHEMATICAL FOCUS

Collections can be made equal by adding or subtracting.

PEDAGOGICAL FOCUS

The tasks in this suggestion have multiple possible answers that students can create using physical models if needed. Some students might count and compare, while others can explore and exhaust possibilities.

It is suggested that you do 'Marbles (1)' as a full-class experience using balance scales, but do not require students to record responses individually. If students are unfamiliar with scales or a coat hanger with paper clips, they may first need opportunities to familiarise themselves with the materials.

In the latter tasks, you might prefer that students create their own drawings, or you could provide prepared diagrams on which students can record their solutions. Refer to **BLM 3: Marbles (1)** for the main task, and **BLM 4: Marbles (2)** and **BLM 5: Marbles (3)** when consolidating the learning. One possibility is to use different coloured balls in each container to assist student thinking when counting.

MARBLES **1**

How can we make the containers of marbles balance without removing any marbles from the left-hand container?

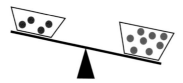

Give as many answers as you can.

Comment

A variety of solutions can be found without removing marbles from the left-hand container. Two marbles could be removed from the right-hand container and added to the left-hand container, leaving 6 marbles in each container. Or 4 marbles could be removed from the right-hand container so there are 4 marbles in each container. It is possible to add new marbles to the containers, such as 4 marbles added to the left container to have 8 marbles in each container.

CONSOLIDATING THE LEARNING

There are two further tasks that explore the idea of making collections equal and consolidate the learning from the first task.

MARBLES ❷

How can we make the containers of marbles balance without removing any marbles from the left-hand container?

Give as many answers as you can.

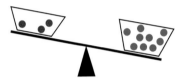

MARBLES ❸

How can we make the containers of marbles balance without removing any marbles from the left-hand container?

Give as many answers as you can.

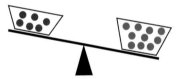

SUGGESTION 3: RECORDING OUR THINKING WITH NUMBER SENTENCES

MATHEMATICAL FOCUS

Number sentences can be used to record how collections on ten-frames were made equal.

PEDAGOGICAL FOCUS

The tasks in this suggestion have multiple possible answers, which students can create using the ten-frames. Students can move to using symbols to record their answers. Students have opportunities to make decisions about an experience they have had (or can see).

This is essentially the same thinking as the earlier suggestions, but provides more structured representation. Encourage students to visualise a possible solution before moving the counters, and to record responses numerically. Additionally, students should move beyond counting individual counters each time, to realise that the top row contains 5 because we have counted it before.

If students are unfamiliar with ten-frames, they may need to familiarise themselves with the materials so they can use them with mathematical purpose. It is helpful to use different coloured counters on each frame to assist student thinking when counting, as well as moving towards recording responses numerically.

Have discussions about students' numerical recordings using questions such as, *What is the same about the number sentences? What is different about them?* Such questions can assist students to notice the relationship between the different equations. For example, that 5 + 2 = 7 and 9 − 2 = 7 are equivalent ways of expressing 7. It is expected that over time students will develop an understanding that the equals sign symbolises the relationship between two equal quantities. This understanding is explored fully in Suggestions 4 and 5.

TEN-FRAMES ①

Make the number of counters on these frames the same (without removing any from the top row of the ten-frame on the left). Do this as many different ways as you can.

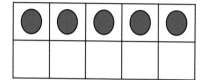 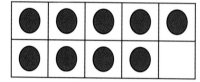

Comment

The multiple possible answers provide students with opportunities to compare the different types of equations (e.g. 9 = 5 + 4, 9 − 4 = 5). For example, the solution found by moving two counters from the right-hand frame to the left can be recorded as: 9 − 2 = 5 + 2.

CONSOLIDATING THE LEARNING

There are three further tasks that explore the idea of making collections equal using a ten-frame. The first two consolidate the learning from the first task. The last task extends the learning by moving beyond 10.

TEN-FRAMES ②

Make the number of counters on these frames the same, without removing any from the top row of the ten-frame on the left. Do this in as many ways as you can.

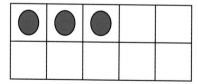 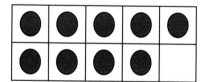

TEN-FRAMES ③

Make the number of counters on these frames the same, without removing any from the top row of the ten-frame on the left. Do this in as many ways as you can.

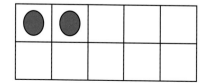

OXFORD UNIVERSITY PRESS

TEN-FRAMES 4

Make the number of counters on these frames the same, without removing any from the ten-frame on the left. Do this in as many ways as you can.

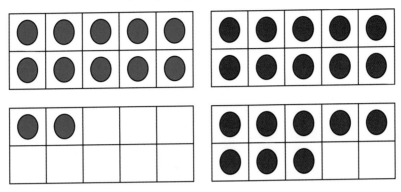

SUGGESTION 4: BALANCING EQUATIONS

MATHEMATICAL FOCUS

Both sides of an equation need to be equivalent when the equals sign is used.

PEDAGOGICAL FOCUS

Students are moving towards using symbols to explore equivalence in a more abstract way. Students should use skills and knowledge they have learned in the previous suggestions. Some students, especially in Foundation, may need to model this, such as with counters.

The latter suggestions use a '?' to indicate an unknown. You might like to use a symbol such as '•' to represent the unknown. Many students are ready to use letters (as in algebra). Where there are two unknowns, you could use a different symbol or letter for the other unknown.

It is important to encourage students to regularly check what they write, and to correct mistakes if necessary; to think about how to make calculations easier; and to develop an understanding of the equals sign.

MAKING IT RIGHT 1

I saw a student write this equation and then change just one number to make it correct:

$$5 + 4 = 11$$

What might the student have done? Can you do it in other ways?

Comment

Changing one number to make it correct, the 5 could be changed to make $7 + 4 = 11$ or the 4 could be changed to make $5 + 6 = 11$. Alternatively, the 11 could be changed to make $5 + 4 = 9$.

CONSOLIDATING THE LEARNING

There are five further tasks that explore the idea of making two sides of a number sentence equal. The first task consolidates the learning. The other four tasks extend that learning by opening up the task to more possibilities and more complex numbers. It is expected that most students can find multiple solutions.

MAKING IT RIGHT ②

I saw a student write this equation and then change just one number to make it correct:

$$4 + 6 = 9$$

What might the student have done? Can you do it in other ways?

CREATING EQUATIONS ①

$$4 + \bullet = \blacksquare$$

What are some numbers to put in place of the symbols to make the equation correct?

CREATING EQUATIONS ②

$$a + 5 = b$$

What are some numbers to put in place of the letters to make the equation correct?

CREATING EQUATIONS ③

$$4 + c = 2 + d$$

What are some numbers to put in place of the letters to make the equation correct?

CREATING EQUATIONS ④

$$e + 1 = 4 + f$$

What are some numbers to put in place of the letters to make the equation correct?

SUGGESTION 5: BALANCING EQUATIONS BEYOND 20

MATHEMATICAL FOCUS

There is a relationship between the numbers on either side of the equals sign.

PEDAGOGICAL FOCUS

Students are moving to using symbols to explore equivalence in a more abstract way. They should identify the missing number in the equations by looking at the relationship between the numbers on either side of the equals sign, rather than by calculating the sum of the expression on the left-hand side. Encourage students to draw on their knowledge of partitioning, such as

OXFORD UNIVERSITY PRESS

5 into 2 + 3, which is a useful strategy for intuitive and mental approaches to some calculations. Teachers will spend time in the 'Summarise' phase of the lesson on the different ways students worked out the answers. Questions such as the following might assist some students to see the relationship between the expression: *What do you notice about the numbers 18 and 20? How might knowing that the difference between these numbers is 2 help you find a solution?*

THE MISSING NUMBER ①

Without writing anything, work out what is the missing number in this equation:

$$18 + 5 = 20 + ?$$

What advice would you give to someone on how to work out answers to questions like that in their head?

Comment

Initially, students might add the 5 to 18 to work out the missing number. This can be addressed by making the numbers more complex.

Some students might respond to a linear representation of the problem, such as:

18	5
20	?

CONSOLIDATING THE LEARNING

There are three further tasks that explore the idea of making two sides of a number sentence equal. The first one consolidates the learning from the initial task. The other two tasks extend the learning by comparing and contrasting the equations and exploring larger numbers. It is expected that most students will be able to find multiple solutions. Teachers might notice some students working systematically to find a range of solutions.

THE MISSING NUMBER ②

Without writing anything, work out what is the missing number in these equations:

$$29 + 6 = 30 + ?$$
$$19 + 8 = 20 + ?$$

What advice would you give to someone about how to work out the answers to questions like that in their head?

SAME AND DIFFERENT

Without writing anything, work out what the missing number is in these equations:

$$57 + 6 = 60 + ?$$
$$57 + ? = 60 + 3$$
$$? + 4 = 60 + 3$$

What is the same and what is different about the equations?

What might be the missing numbers in this equation?

Give as many different possibilities as you can.

$$15 + a = 20 + b$$

SUGGESTION 6: PRINCIPLES OF ADDITION AND SUBTRACTION

MATHEMATICAL FOCUS

Patterns can be used to make sense of addition and subtraction.

PEDAGOGICAL FOCUS

This suggestion is more formal and is suitable for Year 2 students and above. It seeks to go beyond procedures for addition and subtraction by encouraging students to look for patterns. For example, students might come to realise that the second missing number is 6, and that the third missing digit is one more than the first.

A horizontal form of the addition equation with numbers missing is used. The digit represented by __ can be the same or different.

MISSING NUMBER ADDITION

I did an addition question correctly for homework, but my printer ran out of ink.

I remember it looked like:

$$\_\,4 + \_\ = \_\,0$$

What might be the digits that did not print?

Give as many sets of answers as you can.

Comment

Some possible answers could be 14 + 6 = 20, 24 + 6 = 30, 34 + 6 = 40, building up by 10s until 84 + 6 = 90.

CONSOLIDATING THE LEARNING

There are three further tasks in this suggestion that explore patterns. The first task consolidates the learning; the other two tasks extend that learning by comparing and contrasting addition equations and subtraction.

MORE MISSING NUMBER ADDITION

I did an addition question correctly for homework, but my printer ran out of ink. I remember it looked like:

$$\underline{}\,9 + \underline{} = \underline{}\,0$$

What might be the digits that did not print?

Give as many sets of answers as you can.

ADDITION PATTERNS

Work out some possible missing numbers in each of these equations.

$$\underline{}\,4 + \underline{} = \underline{}\,0$$
$$\underline{}\,4 + \underline{} = \underline{}\,2$$
$$\underline{}\,2 + \underline{} = \underline{}\,4$$
$$\underline{}\,0 + \underline{} = \underline{}\,4$$

What is the same and what is different about the sentences?

MISSING NUMBER SUBTRACTION

I did a subtraction question correctly for homework, but my printer ran out of ink. I remember it looked like:

$$3\,\underline{} - 2\,\underline{} = \underline{}\,2$$

What might be the digits that did not print?

Give as many answers as you can.

CHAPTER 4

COUNTING PATTERNS

OVERVIEW

This sequence is suitable for students in Years 1 and 2. However, some tasks could be adapted for Foundation students who have the prerequisite number knowledge and counting skills. As this sequence builds on prior learning from the 'Counting principles' sequence and explores patterns and number relationships, it is recommended that students have completed or are familiar with the big ideas underpinning the 'Counting principles' sequence.

The first two suggestions predominantly focus on counting collections, whereas the others explore number patterns, initially on a number chart and then without one, as students work towards demonstrating sophisticated strategies with counting patterns.

The following is a summary of the suggestions in this sequence.

	MATHEMATICAL FOCUS
Skip-counting a collection of objects	Skip-counting is a counting strategy used to determine the number of objects in a collection.
Skip-counting an abstracted collection of objects	Skip-counting can be used to determine the total in a collection when all objects are not visible.
Skip-counting patterns on a number chart from zero	A number chart is a useful tool when exploring the relationship between different counting patterns.
Skip-counting patterns on a number chart from any number	There are connections between different skip-counting patterns when starting from any number.
Skip-counting patterns (choral counting)	A number chart can be constructed in various ways so that different counting patterns emerge.
Exploring advanced counting patterns	Advanced counting patterns require the use of sophisticated strategies beyond skip-counting.

RATIONALE

The goal of this sequence is to extend students' understanding of counting numbers by examining number relationships, to build place-value understanding and computational skills (including fluency), and to provide opportunities for reasoning and problem solving.

Initially, students explore these ideas using concrete materials, then number charts and choral counting. Choral counting is introduced in the latter sequences to provide students with opportunities to explore in-depth mathematical ideas related to pattern and structure as they predict and notice patterns and relationships.

The following suggestions aim to complement teachers' usual approaches for exploring counting patterns with their students. Note the critical importance of the order in which the tasks occur in the sequence and the follow-up consolidating tasks within each suggestion. Some tasks are differentiated with enabling and extending prompts. To extend students, encourage them to form generalisations about the relationship between the number patterns.

LANGUAGE

skip-counting, objects, collection, counters, number, count, altogether, number chart, pattern, record, backwards, forwards, doubled

LAUNCHING THE TASKS

For Suggestions 1 and 2, you could prepare students by revisiting subitising activities involving multiple ten-frames from the 'Counting principles' sequence, as well as skip-counting by 2s. Beyond this, the whole class could be engaged in oral counting activities involving body percussion whereby students clap on every second (odd or even) count. Musical instruments (triangles or a drum) could be introduced to increase student engagement in counting patterns. For Suggestions 3 and 4, you could prepare some number charts (e.g. 0 to 109 or 119) with different starting and ending numbers for students to identify and explain any pattern they see. For Suggestion 5, large sheets of paper (e.g. butcher's paper) are needed to record whole-class choral counting.

ASSESSMENT

Suggested pre- and post-assessment task: Suggestion 3: 'My friend'. Use different numbers in the post-assessment task.

The following are some specific statements to inform assessment.
Students are learning about skip-counting when they can:

- use and explain non-count-by-one strategies (including mental groupings) to efficiently determine the number of visible items in a large collection
- use, represent and explain non-count-by-one strategies (including mental groupings) to efficiently determine the number of non-visible items in a large collection
- explore, identify and explain forward and backward skip-counting patterns by 2s, 3s, 5s and 10s on a number chart starting at zero
- compare and explain relationships between various skip-counting sequences involving numbers up to and including 10, using number charts and starting at zero
- explore, identify and explain forward and backward skip-counting patterns involving numbers up to and including 10, using number charts and starting at any number
- compare and explain relationships between skip-counts starting from zero and skip-counting starting from a non-zero number
- explore, describe and justify different forward and backward skip-counting patterns on a class-constructed number chart
- identify and describe relationships between various forward and backward skip-counting patterns on a class-constructed number chart
- explore, identify, describe and justify counting patterns using strategies beyond skip-counting
- organise and record their thinking about counting patterns in systematic ways.

MATHEMATICAL FOCUS

Skip-counting is a counting strategy used to determine the number of objects in a collection.

PEDAGOGICAL FOCUS

BLM 6

Oxford
OWL

Skip-counting is a counting strategy. Students are encouraged to calculate and skip-count the number of objects presented in the images in any manner *not* involving counting by 1s. All images (on **BLM 6: Counters (1)**) lend themselves to counting by 2s, and this is generally how most students (5–7 year olds) will approach the problem initially. The follow-up question asking students to do it 'another way' is critical and is designed to push students to start thinking of different, efficient means of skip-counting and counting collections. Operating with numbers and working flexibly with what the counting unit might be (e.g. 2s, 4s) is a precursor to multiplicative thinking, which this suggestion begins to develop.

COUNTERS 1

Can you work out how many counters there are in the picture, without counting by 1s?
Can you do it another way?
Which do you think is the quickest and easiest way to count them? Why?

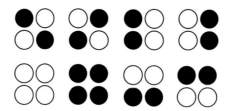

Comment

For this task, some students may wish to count the different coloured dots and add them together, or imagine the dots being organised into ten-frames to assist with enumerating the set. Consequently, as well as exploring counting patterns, tasks such as these encourage students to consider additional structural relationships (e.g. counting in multiples, doubling, place value).

Enabling prompt

Can you work out how many counters there are in the picture, without counting by 1s?

Extending prompt

How many different rectangular arrays can you create from this collection of counters?

Draw your solutions and label your rectangles.

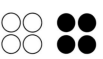

CONSOLIDATING THE LEARNING

BLM 7

Oxford
OWL

There are three tasks that consolidate the learning from the first task. Enabling and extending prompts are provided for some tasks, but because most have a low floor and high ceiling and/or multiple solutions, prompts may not be needed. Give students a copy of **BLM 7: Counters (2)**, or project the picture onto a screen for the students to see.

COUNTERS ❷

Work out how many counters there are in the picture without counting by 1s.

Can you do it another way? Which do you think is the quickest and easiest way to count these counters? Why?

ANTS

Below is a small army of ants. Work out how many ant legs are in the picture without counting by 1s.

Can you do it another way? Which do you think is the quickest and easiest way to count the ant legs? Why?

Enabling prompt

To the right is a very small army of ants. Work out how many ant legs are in the picture without counting by 1s.

Can you do it another way? Which do you think is the quickest and easiest way to count the ant legs?

Extending prompt

What if there were 20 ants and their leader, Maximus, in the ant army? How many legs would there be altogether?

Can you do it another way? Which do you think is the quickest and easiest way to count the ant legs?

SPIDERS

Below is a large family of spiders. Work out how many spider legs are in the picture without counting by 1s.

Can you do it another way? Which do you think is the quickest and easiest way to count the spider legs? Why?

Enabling prompt

Below is a small family of spiders. Can you work out how many spider legs are in the picture without counting by 1s?

Can you do it another way? Which do you think is the quickest and easiest way to count the spider legs? Why?

Extending prompt

How many spider legs in a family of 16 spiders? Can you do it another way? Which do you think is the quickest and easiest way to count the spider legs? Why?

SUGGESTION 2: SKIP-COUNTING AN ABSTRACTED COLLECTION OF OBJECTS

MATHEMATICAL FOCUS

Skip-counting can be used to determine the total in a collection when all objects are not visible.

PEDAGOGICAL FOCUS

In addition to building on the first pedagogical focus of requiring students to think of different efficient means of counting a collection of objects, this task requires students to imagine the objects (as they are not presented concretely) and represent them in some way. Requiring students to represent their thinking appropriately is critical for empowering them to solve a variety of worded problems.

HOW MANY FEET?

How many feet are in the room right now?
How did you work it out? Show me.
Can you find another way of counting them?
Which do you think is the quickest and easiest way to count them? Why?

Comment

Students might start counting the actual feet of students in the class by ones, but unless there is some order in the way students' feet are arranged, this method is prone to error. Students might know a suitable strategy to count the number of students and then to double the count but may not have the mathematical skills to calculate the answer.

OXFORD UNIVERSITY PRESS

Enabling prompts

Draw your family. How many shoes are needed for your family?

How many shoes are there in the illustration?

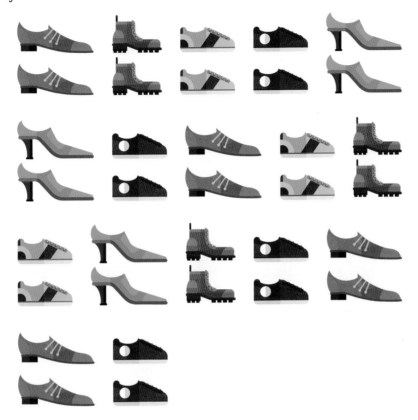

Extending prompts

How many hands and feet are in the room right now? Can you find another way of counting them?

If there are 68 hands and feet in the room right now, how many people are there?

CONSOLIDATING THE LEARNING

Two further tasks consolidate the learning from the above task.

HOW MANY FINGERS?

How many fingers are in the room right now? How did you work it out? Show me.

Can you find another way of counting them? Which do you think is the quickest and easiest way to count them? Why?

Enabling prompts

Draw your family. How many fingers are there altogether in your family?

How many fingers are there in the illustration below?

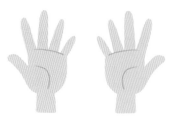

Extending prompt

I forgot to tell you, thumbs don't count as fingers …

FINGERS AND TOES

Read *Ten Little Fingers and Ten Little Toes* by Mem Fox and Helen Oxenbury.

How many little fingers and toes are counted in the book altogether? How did you work it out? Show me. Can you find another way of counting them?

Which do you think is the quickest and easiest way to count them? Why?

Enabling prompt

Draw your family. How many fingers are there altogether in your family?

Extending prompt

I forgot to tell you, thumbs don't count as fingers …

SUGGESTION 3: SKIP-COUNTING PATTERNS ON A NUMBER CHART FROM ZERO

MATHEMATICAL FOCUS

A number chart is a useful tool when exploring the relationship between different counting patterns.

PEDAGOGICAL FOCUS

The focus for this suggestion is on getting students to examine how the various counting sequences are related to one another. Providing the number chart representation is intended to support the exploration of these relationships. Use number charts (e.g. from 0 to 109 or 119) that are large enough for students to place coloured counters on.

The enabling prompt with the number charts that include shaded counting patterns should be used for all suggestions in this sequence, so students can become familiar with these patterns and with what the prompt represents. Some tasks have a range of possible answers, allowing students to make their own choices, count backwards and forwards from a variety of positions within a counting sequence, and explore the relationship between patterns.

Prior to launching the task, the whole class could play a game of 'I spy' to explore, identify and explain any 'pattern' students see on the number chart. Alternatively, students can place their counters above the tens column, 'where the zero would be'. Teachers may also consider using an interactive hundreds chart during the whole-class discussion phase of the lesson and/or to launch the task (e.g. 'Splat squares' on the UK website Primary Games, https://www.primarygames.co.uk/SplatSquares_WebGL/index.html).

EXPLORING NUMBER CHARTS (THIRD TIME LUCKY)

Starting at 0, I skip-counted by 2s to 100,

placing a counter on all the numbers I landed on.

Next, I skip-counted by 5s to 100,

again placing a counter on all the numbers I landed on.

Finally, I skip-counted by 10s to 100,

again placing a counter on all the numbers I landed on.

What are the numbers with three counters on them – the numbers I landed on three times?

Enabling prompt

Starting at 0, I skip-counted by 2s to 20, placing a counter on all the numbers I landed on. Next, I skip-counted by 5s to 20, again placing a counter on all the numbers I landed on. What are the numbers with two counters on them – the numbers I landed on twice?

NUMBER PATTERNS

What number patterns can you see?

Extending prompts

If I continued skip-counting to 200 (by 2s, 5s and 10s), how many numbers would I land on three times?

What if I continued counting to 780? What about 1000?

Describe the pattern for working out how many numbers I have landed on three times.

Comment

You might also discuss or ask students to predict where the next counter or other counters will go once a pattern begins to appear. Then check their prediction.

CONSOLIDATING THE LEARNING

Five further tasks consolidate the learning from the above task.

1	2	3	4	5	6	7	8	9	10
11	12	13	14	15	16	17	18	19	20
21	22	23	24	25	26	27	28	29	30
31	32	33	34	35	36	37	38	39	40
41	42	43	44	45	46	47	48	49	50

1	2	3	4	5	6	7	8	9	10
11	12	13	14	15	16	17	18	19	20
21	22	23	24	25	26	27	28	29	30
31	32	33	34	35	36	37	38	39	40
41	42	43	44	45	46	47	48	49	50

1	2	3	4	5	6	7	8	9	10
11	12	13	14	15	16	17	18	19	20
21	22	23	24	25	26	27	28	29	30
31	32	33	34	35	36	37	38	39	40
41	42	43	44	45	46	47	48	49	50

1	2	3	4	5	6	7	8	9	10
11	12	13	14	15	16	17	18	19	20
21	22	23	24	25	26	27	28	29	30
31	32	33	34	35	36	37	38	39	40
41	42	43	44	45	46	47	48	49	50

1	2	3	4	5	6	7	8	9	10
11	12	13	14	15	16	17	18	19	20
21	22	23	24	25	26	27	28	29	30
31	32	33	34	35	36	37	38	39	40
41	42	43	44	45	46	47	48	49	50

EXPLORING NUMBER CHARTS (FOURTH TIME LUCKY)

Starting at 0, I skip-counted by 2s to 50, placing a counter on all numbers I landed on.

Next, I skip-counted by 3s to 50, again placing a counter on all numbers I landed on.

After that, I did the same thing counting by 5s, and then 10s.

There is only one number with four counters on it. What is that number?

Extending prompt

What if I continued skip-counting to 100 instead of 50? How many numbers would I have landed on four times? What are these numbers?

List all the numbers I would land on four times if I continued counting to 1000.

Do you notice any interesting patterns with these numbers?

EXPLORING NUMBER CHARTS (WHICH NUMBERS WILL SURVIVE?)

Starting at 0, I skip-counted by 2s to 40, crossing off the numbers as I went. Then I did the same thing, but instead skip-counted by 3s. Next, I did it by 4s. Finally, I skip-counted again, but counted by 5s.

Some numbers were crossed off more than once, but some numbers survived – they weren't crossed off at all. Can you guess which 10 numbers survived? Write down the numbers. Now check if you are right.

Extending prompt

What if I also skip counted by 6s, 7s, 8s, 9s and 10s?

Would all 10 numbers still survive?

How many more numbers would get crossed off?

LUCKY DICE ❶

My dad offered me a deal. I choose any number on a hundreds chart. He'll then roll a 6-sided dice, and we'll count by whatever number he rolled (from zero). If we land on my number, he'll give me 10 dollars. If we skip my number, I'll give him 10 dollars.

What are some good numbers I could choose?

Should I take the deal?

Extending prompt

Prove that there is a number on the hundreds chart that will *guarantee* me winning the deal every single time.

What if he rolled a 10-sided dice instead, but let me choose any number up to 1000? How would this change the number I might choose? Am I still guaranteed of winning the deal?

MY FRIEND ❶

My friend told me three number sequences.

The first sequence had the number 24 in it. What was she counting by? Do this a different way. What do you notice between the different patterns you have made?

The second sequence had the number 36 in it. What was she counting by? Do this a different way.

The third sequence had the number 50 in it. What was she counting by? Do this a different way.

Combine all three patterns. What do you notice about the patterns? Are there any numbers that are counted more than once?

SUGGESTION 4: SKIP-COUNTING PATTERNS ON A NUMBER CHART FROM ANY NUMBER

MATHEMATICAL FOCUS

There are connections between different skip-counting patterns when starting from any number.

PEDAGOGICAL FOCUS

This suggestion builds on the previous pedagogical focus. The additional focus is getting students to consider the relationship between skip-counting from zero and skip-counting from a non-zero starting point; specifically, the idea that starting from a non-zero sequence simply shifts the entire pattern over (in algebraic terms, the sequence has been translated).

Prior to launching the task, students could predict then confirm how patterns might change when a different number chart or starting number is used.

EXPLORING NUMBER CHARTS FROM ANY NUMBER

Starting at 6, I skip-counted by 2s to 100, placing a counter on all numbers I landed on.

Next, again starting at 6, I skip-counted by 3s to 100, placing a counter on all numbers I landed on. After that, I did the same thing counting by 5s, and then 10s.

There are three numbers with four counters on them. What are those numbers?

Comment

This task has a range of possible answers allowing students to make their own choices, count backwards and forwards from a variety of positions within a counting sequence, and explore the relationship between patterns.

Enabling prompt

Starting at 6, I skip-counted by 2s to 26, placing a counter on all the numbers I landed on. Next, again starting at 6, I skip-counted by 5s to 26, again placing a counter on all the numbers I landed on. What are the numbers with two counters on them – the numbers I landed on twice?

Extending prompt

What if I continued skip-counting to 1000? How many numbers would I have landed on four times? What are these numbers? What do you notice about the patterns with these numbers?

CONSOLIDATING THE LEARNING

Two further tasks consolidate the learning from the above task.

LUCKY DICE 2

My dad offered me another deal. Again, I can choose any number on a hundreds chart. He will roll a 6-sided dice, and we'll count by whatever number he rolls. But this time, we'll start counting from 11. If we land on my number, he'll give me 10 dollars. If we skip my number, I'll give him 10 dollars.

What are some good numbers I could choose? Should I take the deal?

MY FRIEND 2

My friend told me a number sequence. She started counting at 7, and the sequence had the number 27 in it. What was she counting by? Do this a different way.

What do you notice comparing the different patterns you have made?

My friend told me another number sequence that had the numbers 50 and 56 both in it. What did my sequence look like? Do this a different way.

SUGGESTION 5: SKIP-COUNTING PATTERNS (CHORAL COUNTING)

MATHEMATICAL FOCUS

A number chart can be constructed in various ways so that different counting patterns emerge.

PEDAGOGICAL FOCUS

It is assumed a number chart will be created in real time as the teacher records the choral count. In this way, a number chart is created as an artefact of the choral count, rather than as a representation to support student thinking. The count is stopped at strategic points to allow students the opportunity to notice patterns and predict certain numbers. The patterns generated by the teacher in recording the impromptu number chart are then discussed. Ideally, the teacher keeps the choral count going until the class has reached a number greater than 110. The discussion about the patterns that students notice should be in-depth. As this is a whole-class task, there are no enabling or extending prompts.

Consider conducting the choral counts at least twice, recording the numbers in a different way, so different patterns emerge. It is important to leave enough space on the recording sheets/board to document the patterns that students notice.

Hint: Do the task yourself before teaching, to consider the patterns the students might notice. To assist with the alignment of numbers, it is also helpful to lightly pencil in rows and columns on the large sheets of paper.

One way of organising the count:

90	92	94	96	98
100	102	104	106	108
110	112	114	116	118

Another way of organising the count:

80	84	88	92	96	100
82	86	90	94	98	102

Prior to launching the task, prepare the large sheet of paper/board with appropriately sized rows and columns lightly pencilled in. Students' thinking could be tuned in to this task via a whole-class oral skip-count by 2s activity starting from a familiar number. The actual launch includes posing the task and encouraging students to think about what numbers might come next before you start the count.

CHORAL COUNTING BY 2s

As a class, we are going to begin counting by 2s from 80.

I will record our count.

Let's count by 2s from 80 again, but this time I am going to record our counting a different way. What do you notice?

Comment

Once you have finished recording the count, ask students, *What do you notice?* Use different coloured pens to highlight the patterns and relationships students describe and justify in a similar way to that produced by a Year 1 class shown below. Students could use these patterns to predict the next number(s) in a row, column or diagonal line of numbers.

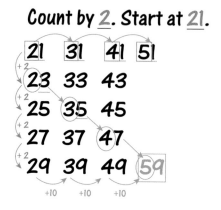

Count by _2_. Start at _21_.

CONSOLIDATING THE LEARNING

Three further tasks consolidate the learning from the above task.

CHORAL COUNTING BY 5s

As a class, we are going to begin counting by 5s from 21 and see how far we can go. I will record our count. What do you notice?

Let's count by 5s from 21 again, but this time I am going to record our counting a different way. What do you notice?

CHORAL COUNTING BY 4s

As a class, we are going to begin counting by 4s from 20 and see how far we can go. I will record our count. What do you notice?

Let's count by 4s from 20 again, but this time I am going to record our counting a different way. What do you notice?

CHORAL COUNTING BY 10s

As a class, we are going to count by 10s backwards from 346. What do you notice?

Let's count by 10s backwards from 346 again. What do you notice?

SUGGESTION 6: EXPLORING ADVANCED COUNTING PATTERNS

MATHEMATICAL FOCUS

Advanced counting patterns require the use of sophisticated strategies beyond skip-counting.

PEDAGOGICAL FOCUS

The tasks in this suggestion are more challenging and are intended to extend students' thinking by requiring them to engage with more complex patterns. The important teaching point is that sometimes it is insufficient to consider only two successive terms in a sequence to discern a pattern; instead, the student needs to examine the relationship between several successive terms to work out the rule. For example, in the pattern 2, 4, 8, 16, considering only the first two terms might lead a student to conclude that the pattern involves counting by 2s (or + 2). It is only through examining the relationships between several consecutive terms (2 → 4; 4 → 8; 8 → 16) that the student realises that the pattern involves doubling, or × 2.

Encourage students to record and organise their thinking in systematic ways, including a table. Ask students to explain and justify their answers by referring to the relationships between several terms in the pattern. Discuss and share the different responses and strategies for solving the task.

MAGIC DOUGHNUT TREE 1

Kai was having his birthday party on Friday, so the family decided to not pick any of the doughnuts off the tree until then. On Monday, there were 3 doughnuts on the tree.
How many doughnuts were on the tree in time for Kai's party on Friday?

Comment

Prior to launching the task, you are encouraged to create a story to which the students can relate as a way of posing the task. For instance, the story of the Magic Doughnut Tree starts:

> Kai and Amaya loved doughnuts, so their mum decided to plant a doughnut tree. The tree was magical. Every day, the number of doughnuts on the tree doubled. If there were 2 doughnuts on the tree today, there would be 4 doughnuts on the tree tomorrow, and 8 doughnuts on the tree the day after that.

Enabling prompt

What if there was only one doughnut on the tree on Monday? Each day the number of doughnuts double. How many doughnuts would be on the tree in time for Kai's party on Friday?

Extending prompt

How many doughnuts would be on the double doughnut tree if Kai decided to have the party on Saturday instead? What about if he had the party on Sunday? Can you keep the pattern going?

CONSOLIDATING THE LEARNING

Four further tasks consolidate the learning from the above task.

MAGIC DOUGHNUT TREE ❷

After Kai's party on Friday, there are 5 doughnuts left on the tree on Saturday morning (Kai was kind enough to not pick all the doughnuts!). Amaya decides to try and earn some money selling doughnuts, and her mum agrees to let her use the doughnut tree for a week. After one week is up, Amaya will sell whatever doughnuts she can pick off the tree to her classmates for one dollar each.

Later that day, her cousin Max offers her $100 if he can have the tree for a week instead. Should Amaya take the deal? Explain your thinking.

Enabling prompt

Use this table to help you work out whether Amaya should make the deal with Max

DAY	DOUGHNUTS
Saturday	5
Sunday	10
Monday	20
Tuesday	
Wednesday	
Thursday	
Friday	
Saturday (picking day)	

Extending prompt

What if Amaya's mum let her have the doughnut tree for 2 weeks? How much money would she make? Use a calculator to help you.

Extra challenge: How long would she need to have the tree for to make one million dollars?

MAGIC LOLLIPOP TREE

Lola and Leo loved lollipops, so their dad decided to plant a lollipop tree. The tree was magical. On Monday, the tree grew one lollipop, on Tuesday it grew two more lollipops, on Wednesday it grew three more lollipops. And the pattern continued.

Lola wants to take some lollipops to school to share. Including Lola, there are 53 children in Foundation class altogether, plus two teachers. How many days does Lola need to wait until she has enough lollipops to share?

Enabling prompt

Use this table to help you work out what day Lola might share her lollipops.

DAY	LOLLIPOPS
1	1
2	1 + 2
3	1 + 2 + 3
4	1 + 2 + 3 + 4

Extending prompt

If Lola decides to give a lollipop to every child in your school, how many days will she need to wait?

THE POCKET MONEY PROBLEM

Olive's parents have come to her with a proposal. They will start paying her pocket money at the beginning of next year, but she has to choose between these two options.

Option 1: $20 per month

Option 2: 10 cents in January, and then the amount will double for February, double again for March and continue to double every month until December. At first glance, which option do you think might be better and why?

Work out how much money Olive would get in over one year with each option. Clearly show your working out.

Which option was better, and by how much?

Extending prompt

If the options were to continue for a second year, how much money would Olive have if she chose Option 1? What about Option 2?

Olive has just turned 6 years old. If she chose Option 2, and her parents kept up the deal, how much pocket money would she receive the month she turned 18?

CHAPTER 5

PLACE VALUE

This sequence is suitable for students in Years 1 and 2. It builds on prior learning from the 'Counting principles' sequence, in which there was a focus on composing (putting together) and decomposing (taking apart) the teen numbers. It is therefore recommended that students have completed or are familiar with the big ideas in the 'Counting principles' sequence before being introduced to this sequence. Students are only ready to conceptually understand place value when they can count on from numbers greater than ten.

The first four suggestions predominantly focus on working with numbers to 100, but can be extended to 3-digit numbers for students in Year 2.

The following is a summary of the suggestions in this sequence.

	MATHEMATICAL FOCUS
Counting and representing numbers	Numbers can be represented in different ways. Grouping in tens makes large quantities easier to count.
Comparing and ordering numbers	Numbers can be compared and ordered using place value knowledge.
Exploring place value patterns in a number chart	There are many different patterns within a number chart.
Using the number chart for additive thinking	Number charts can be used to support addition and subtraction.
Using the number chart for additive thinking beyond 100	Patterns in a number chart can support addition and subtraction beyond 100.
Estimating numbers on a number line	A number line is a tool to represent the relative size of a number.

RATIONALE

The goal of this sequence is to develop an understanding of place value that supports a conceptual understanding of addition and subtraction processes. To advance conceptual understanding, it is important to develop student knowledge of key mathematical ideas that are fundamental to place value, such as unitising and equivalence.

This sequence begins by building on the big idea of unitising, i.e. that ten objects means one ten. This idea was introduced in the 'Counting principles' sequence and is central to learning place value. By exposing students to a variety of meaningful contexts in which they organise objects into groups to make them easier to count, students will develop an understanding of this fundamental concept.

OXFORD UNIVERSITY PRESS

Another big idea central to understanding place value is recognising equivalent ways of representing a number. Students often learn the names of place value positions without understanding that the tens place represents how many groups of ten and can help to work out how many. Similarly, students might not be able to explain that ten objects also means one ten, or that 100 objects is the same as 10 groups of ten.

In this sequence the number chart is used to develop an understanding of place value that focuses on whole numbers and quantity value. Learning early mental computation strategies usually involves adding and subtracting initially by 10, and later in multiples of 10, with numbers other than multiples of ten, e.g. 27 and 10 more is 37. Such strategies also involve being flexible in adding and subtracting both tens and ones. Developing a conceptual understanding of place value is critical in developing number sense and computation strategies.

The suggestions in this sequence are intended to be complementary to other experiences that teachers commonly use to reinforce place value concepts. Such experiences may include using ten-frames, using various manipulatives to keep track of the number of days at school or counting the number of days to a special event.

The suggestions in this sequence can be explored multiple times to help students develop flexibility in their thinking about number and the underlying structure of place value.

LANGUAGE

more, less, 10 more, 10 less, more than, greater than, less than, pattern, difference, addition, sum, hundreds, tens, ones, place value, rename

LAUNCHING THE TASKS

In launching the tasks in this sequence, it is important to clarify the language and representation (materials, drawings, symbols) of the tasks. Encourage students to imagine their solution before they start the task.

Tasks included in the 'Counting principles' sequence – 'Directions', 'Clap counting' and 'Dynamic counting' – could be adapted and incorporated into this sequence to support fluency in counting forwards and backwards by 10 from numbers other than multiples of ten.

ASSESSMENT

Suggested pre- and post-assessment task: Suggestion 3: 'Missing numbers on the jigsaw piece'. Use a different number as the known number in the jigsaw for the post-assessment.

The following are some specific statements to inform assessment.

Students are developing their conceptual understanding of place value when they can:

- organise a collection of objects using grouping
- represent and compare quantities using a variety of materials using standard partitioning, and explain their thinking
- represent and compare quantities using a variety of materials using non-standard partitioning
- interpret the greater than sign and its relationship with numbers

- accurately create and compare two-digit numbers with reference to place-value language (i.e. tens and ones)
- recognise patterns in a number chart
- solve problems related to comparing numbers in a number chart
- use appropriate benchmarks to estimate where a number should be placed on a number line
- use appropriate benchmarks to estimate the number that a given point on a number line corresponds to
- approach and check their work systematically and identify patterns to make sure they have found all solutions.

SUGGESTION 1: COUNTING AND REPRESENTING NUMBERS

MATHEMATICAL FOCUS

Numbers can be represented in different ways. Grouping in tens makes large quantities easier to count.

PEDAGOGICAL FOCUS

The focus of this suggestion is developing an understanding of equivalent representations of a number and that groupings of ten make a large quantity easier to count; a central idea to learning place value. It is intended that students will explore and develop flexibility in thinking about numbers in terms of tens and ones.

The connection between numbers and their representations can be too implicit in base ten materials. Other groupable materials that allow students to see 10 ones as ten, or 10 groups of ten as 100 ones (e.g. craft sticks) may be preferable to pre-grouped base ten materials (e.g. MAB). Using money (denominations of $10 and $1) or other base ten materials that are non-proportional is more challenging for students. Some students may have had sufficient experiences with manipulatives and be ready to progress to pictorial representations or recording in equations.

Prior to launching the suggestion, students need to have opportunities to familiarise themselves with materials so they can use them with mathematical purpose. Encourage them to explore the materials before undertaking any formal mathematical work.

When launching the first task, depending on your students, consider providing a large variety of materials and representations, some of which might be pre-grouped (e.g. MAB), some of which students will need to group themselves (e.g. Unifix), and some of which offer a mixture (e.g. bundled and unbundled craft sticks). Include some non-proportional representations in the mix (e.g. money). Learning about representing or partitioning in different ways can be helpful when solving addition and subtraction problems.

The 'Representing numbers' task is intended to develop the idea that we can use groupings of tens and ones to take apart (decompose) a number in standard and non-standard or equivalent ways. For example, 35 can be seen as 30 + 5 (standard partitioning) or 20 + 15 (non-standard partitioning).

OXFORD UNIVERSITY PRESS

No enabling or extending prompts are suggested. Teachers should modify the numbers so they are appropriately challenging for students (e.g. the size of the number could be increased to beyond 100). This task can be done many times using a different target number.

REPRESENTING NUMBERS

Find as many different ways to represent the number 35 as you can.

CONSOLIDATING THE LEARNING

BLM 8

There are three further tasks to consolidate student learning. The first task reinforces the idea that a given quantity can be represented in different ways and has an illustrative worksheet – see **BLM 8: Four representations**.

Oxford OWL

The second and third tasks require students to consider ways to count, group and organise collections that utilise place value to efficiently find totals.

FOUR REPRESENTATIONS

Can you match the cards in the worksheet that show the same number?

HOLD 10

My mum's bus holds a maximum of 10 people.
To get us all to the pool, Mum needed to make 4 trips.
How many people needed to go to the pool?

LARGE NUMBER COUNTING

How many Brainies?
Find two different ways to work out the number of Brainies.

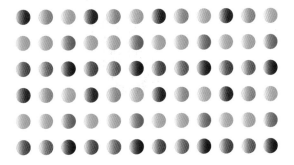

SUGGESTION 2: COMPARING AND ORDERING NUMBERS

MATHEMATICAL FOCUS

Numbers can be compared and ordered using place value knowledge.

PEDAGOGICAL FOCUS

This suggestion is intended to prompt students to use their knowledge of place value to compare and order numbers. The *greater than* sign is introduced early in the suggestion to facilitate a systematic comparison of pairs of numbers. Although the greater than sign is not formally introduced in most official curriculum until later, its early introduction is beneficial on multiple levels. In a place-value context, it introduces students to an efficient means of recording which of two numbers in a set is larger. It can be introduced to serve as a counterpoint to the equals sign (e.g. 7 + 3 > 8 + 1), supporting students to develop a relational understanding of this symbol. That is, an equals sign does not just mean 'the answer', it means that two sides of an equation are equal or balanced (e.g. 7 + 3 = 8 + 2).

Prior to launching the suggestion, explain the meaning of the greater than/less than symbols. Record some simple sentences on the board and ask students to confirm if they are correct, e.g. 8 > 2. The crocodile (or Pacman) metaphor might be useful for helping students to remember which way the sign should face (e.g. the hungry crocodile always eats the larger amount).

For the task 'Which is more?', students are required to understand that the order of the digits is important and that the position of a digit determines which group size it represents. For example, for the numbers 57 and 37 the tens digit determines the order, whereas for 75 and 73 the decision is based on the digits in the ones place. Often students know the names of the positions, e.g. tens place, but do not understand that this represents the number of groups of ten, or make the connection with the quantity. Students may find it helpful to have a template and some number cards to explore the possibilities.

WHICH IS MORE? ①

Place each of these digits onto one of the empty boxes to make the sentence true.

Find as many different ways of doing this as you can.
Record your solutions.

Comment

There are 6 solutions to the 'Which is more?' task: 77 > 35; 77 > 53; 75 > 37; 75 > 73; 73 > 57; 57 > 37. To find all possible solutions (and to be confident that they have found them all), students will need to be systematic in how they approach the problem and how they check that they have found all the possibilities.

OXFORD UNIVERSITY PRESS

Enabling prompt

Using just these number cards, make some numbers that are larger than 10 but smaller than 100.

Extending prompt

My cousin John thought there should be 12 solutions, but I could only find 6. Who is right? Convince me.

CONSOLIDATING THE LEARNING

Four further tasks consolidate the learning. The first task involves different numbers and leads to there being 12 possible answers, rather than 6 (so cousin John would be right this time!), something that students pursuing more systematic approaches to the problem might discover. The remaining tasks all involve tuning students into the size of two-digit numbers to develop an appreciation of place value using different representations/contexts (e.g. digits on cards, numbers in a problem-solving context, numbers on a chart). Depending on where students are at, it is appropriate to modify the tasks so students are exploring three-digit numbers instead (or have the option to do so).

WHICH IS MORE? ❷

Place each of these digits into one of the empty boxes to make the sentence true. Find as many different ways of doing this as you can.

Record your solutions.

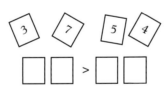

LARGEST AND SMALLEST NUMBERS

Using just these cards, make as many different 2-digit numbers as you can.

What is the largest 2-digit number you can make?

What is the smallest?

THREE TEACHERS

Ms Smith is 56 years old. Ms Chu is 49 years old.

I know that Mr Taylor is older than Ms Chu but younger than Ms Smith.

How old might Mr Taylor be?

Give as many possibilities as you can.

SUGGESTION 3: EXPLORING PLACE VALUE PATTERNS IN A NUMBER CHART

MATHEMATICAL FOCUS

There are many different patterns within a number chart.

PEDAGOGICAL FOCUS

This suggestion involves students exploring patterns within the number chart. There are both vertical and horizontal patterns within the chart. For example, the number one row below another number is always 10 more than that number, when there are ten numbers in each row.

Prior to launching the 'Missing numbers on the jigsaw piece' task, students should be familiar with the number chart and have experience with saying number sequences up to 100, including adding 10 starting from any number, and reading and writing numbers to 100. A useful experience would be to provide the students with a number chart jigsaw to complete. Number chart jigsaws can be created by laminating and cutting up a number chart into jigsaw pieces.

Another idea is to get students to play a 'guess my number' game using a number chart. This activity can be played with the whole class, with either the teacher or a student choosing the number, or in small groups. The intention is that the numbers which have been excluded by an earlier guess are removed. Questions being posed should focus on use of the language 'more' and 'less', for example, *Is your number less than 50?* If the activity works well, this game can be revisited for later suggestions as well.

MISSING NUMBERS ON THE JIGSAW PIECE

What might be the numbers on the L-shaped piece?

I know that one of the numbers is 65.

Give as many possibilities as you can.

OXFORD UNIVERSITY PRESS

Comment

There are 16 possible solutions to this task. For each given orientation of the L, the number 65 could be in four different positions. For example, if the L was horizontal (approximating how it appears in the jigsaw photograph), the four solutions would be:

Alternatively, if the L was oriented so that it looked like the letter L as we would read it on a page, the four solutions would be:

There are also four solutions for each of the remaining two orientations, making 16 solutions in all. The number of possibilities can be increased if you allow the L-shaped piece to be flipped (instead of 16 solutions, there will be 32). To find all possible solutions, students will need to approach the task systematically.

Enabling prompt

What might be the missing numbers on this piece?

Extending prompt

Can you find more than 8 possible combinations?

CONSOLIDATING THE LEARNING

Four tasks consolidate the learning, some of which have multiple possible answers. The 'Letters of the alphabet' task is similar in structure to the original task but is more open-ended. The next task, 'Mistake in the letter', stays with this same theme, but it includes some errors in the number chart pattern for students to identify. The third consolidating task also involves students finding mistakes, that is, instances where the number chart pattern is violated, whilst the final two tasks invite students to play with these same ideas.

LETTERS OF THE ALPHABET

The numbers 62 and 84 are on the same jigsaw piece.
The piece is shaped like a letter of the alphabet.
Draw what that piece might look like and write in the numbers.

MISTAKE IN THE LETTER

Find the mistakes in each letter. Explain how you found them.

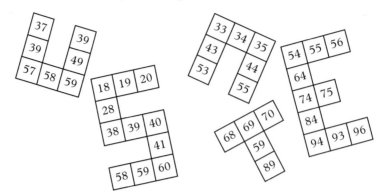

SPOT THE MISTAKE

Find the mistakes in this number chart. Explain how you found them.

1	2	3	4	5	6	7	8	9	10
11	12	13	13	15	16	17	18	19	20
21	22	23	24	25	16	27	28	29	30
31	32	33	34	35	36	37	38	39	40
41	42	33	44	55	46	47	48	49	50
51	52	53	54	55	56	57	48	59	60
61	62	63	64	65	66	67	68	69	70
71	72	73	74	75	76	77	78	79	80
81	82	83	84	85	96	87	88	89	90
91	92	93	94	95	96	97	98	99	100
101	102	103	104	105	106	107	108	109	200

WHAT IS MISSING?

This number chart has not been completed.

Fill in the missing numbers.

1	2	3	4	5	6	7	8	9	10
11	12	13						19	20
21	22	23	24	25	16	27	28	29	30
31	32	33	34						40
41					46	47	48	49	
51					56	57	48	59	
61	62	63	64		66	67	68	69	
71	72				76	77	78	79	80
81	82	83	84			87	88		90
91	92	93	94			97	98		
101	102	103	104		106	107	108		

Which of the following jigsaw pieces could be from our jigsaw, and which are not?
Explain your reasoning.

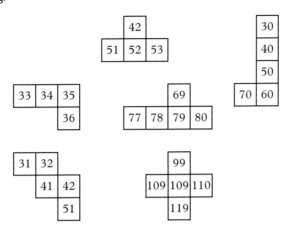

SUGGESTION 4: USING THE NUMBER CHART FOR ADDITIVE THINKING

MATHEMATICAL FOCUS

Number charts can be used to support addition and subtraction.

PEDAGOGICAL FOCUS

Suggestion 4 builds on the two previous suggestions in that it also involves comparing numbers (Suggestion 2) and uses number charts to represent numbers (Suggestion 3). However, it is more sophisticated because it involves students comparing numbers more precisely by looking at the difference between two numbers. The learning focus is on using and interpreting place value. The tasks support students in developing understanding of adding in tens and ones.

Providing students with a printed number chart and counters can make the tasks in the suggestion more accessible. Transparent counters that fit over one square in the number chart are useful to mark the pairs of numbers and still make the patterns visible.

Prior to launching the first task, it is helpful for students to explore how to use the number chart to work out answers to addition and subtraction questions involving tens and ones. For example, *What is 14 + 10? 14 + 8? What is the difference between 24 and 42? How much do you have to add to 18 to get 33?* Students can be invited to consider these problems individually, discuss their approach with a partner, and share their thinking with the class.

WHAT ARE MY TWO NUMBERS?

I am thinking of two numbers on the number chart. One number is 15 more than the other. One of the numbers has a 3 in it. What might my two numbers be?
Give as many answers as you can.

Comment

There are many possible solutions to this task. One set of solutions involves the smaller number having a three in the ones place (numbers such as 3, 13, 23). There are 10 solutions in this set (e.g. 3 and 18; 13 and 28; 23 and 38). Another set of solutions involves the smaller number being in the 30s (e.g. 30 and 45; 31 and 46; 32 and 47). There are 10 solutions in this set, although one of the solutions also belongs to the previous set (i.e. 33 and 48). Alternatively, the larger number might have a 3 in the ones place (e.g. 23 and 8; 33 and 18). There are 9 solutions of this type. Finally, the larger number might be in the 30s (e.g. 30 and 15; 31 and 16). There are 10 solutions in this set, although one of the solutions also belongs to the previous set. There are therefore 37 unique solutions to this task in total. Although it is perhaps unlikely a student would find them all, there are many opportunities for students to be systematic in how they approach the task.

Enabling prompt

I am thinking of two numbers on the number chart.

One number is 5 more than the other.

What might be my two numbers?

Extending prompt

Can you find more than 20 possible answers? Convince me you could find all the possible answers by describing the patterns you can see.

CONSOLIDATING THE LEARNING

Eight further tasks consolidate the learning from the two numbers task. The first consolidating task remains focused on the concept of 'difference', as do the 'Pencils' and 'Eggs' tasks. The final tasks focus on partitioning ('Jumping on a number chart 1, 2') or the notion of 'sum' ('Add to 52', 'Three rows apart', 'Three numbers'), whilst getting students to explicitly attend to structural components of the number chart (e.g. the numbers are two rows apart).

A DIFFERENT TWO NUMBERS

I am thinking of two numbers on the number chart.

The difference between my two numbers is 23.

One of my numbers has a 5 in it.

What might be the numbers?

Give as many answers as you can.

PENCILS

Each box of pencils holds 10 pencils.

I have 4 full boxes and some extra pencils.

My friend had 11 more pencils than me.

How many boxes and how many extra pencils might my friend have?

EGGS

Some egg cartons hold 10 eggs.

Amy has some full cartons and some loose eggs.

Becky has 6 full cartons and some loose eggs.

Amy has 10 fewer eggs than Becky. How many eggs might Amy and Becky have?

JUMPING ON A NUMBER CHART **1**

On a number chart, go from 23 to 55 in four jumps. For each jump, you can jump either horizontally or vertically along the chart.

JUMPING ON A NUMBER CHART **2**

On a number chart, go from 38 to 83 in three jumps. For each jump, you can jump either horizontally or vertically along the chart.

ADD TO 52

I am thinking of two numbers on the number chart.

My numbers are two rows apart. The sum of my numbers is 52.

What might my numbers be? Give as many answers as you can.

THREE ROWS APART

I am thinking of two numbers on the number chart.

The sum of my numbers is 68.

My numbers are three rows apart.

What might my two numbers be?

Give as many answers as you can.

THREE NUMBERS

I am thinking of three numbers on the number chart.

The sum of my numbers is 90.

My numbers are each two rows apart.

What might my three numbers be?

Give as many answers as you can.

MATHEMATICAL FOCUS

Patterns in a number chart can support addition and subtraction beyond 100.

PEDAGOGICAL FOCUS

This suggestion builds on the previous one. It is intended to prompt students to apply their knowledge of place value to solve addition and subtraction tasks with numbers beyond 100.

Providing students with a printed number chart and counters can make the task more accessible. Transparent counters that fit over one square in the number chart are useful to mark the pairs of numbers and make the patterns visible.

If you think students might require more support for this set of tasks, prior to launching the first task you could ask students to study the incomplete number chart and complete it with a partner (optional). However, this reduces the extent to which students will need to visualise the patterns to complete the task. The advantage is that it makes the connection with their previous learning more explicit.

TWO NUMBERS

I am thinking of two numbers on this number chart. The chart is incomplete.

201	202	203	204	205	206	207	208	209	210
211									220
221									230
231									240
241									250
251									260
261									270
271									280
281									290
291	292	293	294	295	296	297	298	299	300

The difference between my numbers is 12.

One of my numbers has a 5 in it.

What might be the numbers?

Comment

The 'Two numbers' task is very similar in structure to the 'What are my two numbers?' task in Suggestion 4. The major difference is that this chart covers the numbers 201 to 300, and some of the numbers on the chart are missing. The sets of solutions to the task are similar. There are 34 unique solutions in total.

The smaller number has a 5 in the ones column: 9 solutions (205 and 217 ... 285 and 297). The smaller number has a 5 in the tens column: 10 solutions (250 and 262 ... 259 and 271) – we already have one solution from the previous set (255 and 267).

The larger number has a 5 in the ones column: 9 solutions (215 and 203 ... 295 and 283). The larger number has a 5 in the tens column: 10 solutions (250 and 238 ... 259 and 247) – although we already have one solution from the previous set (255 and 243).

There are also two additional double-ups across the set of 'smaller numbers with a 5' and 'larger numbers with a 5' (245 and 257; 253 and 265). There are 38 solutions in total. Accounting for those double-ups from the previous set, we are left with 34 unique solutions.

Enabling prompt

Can you write the numbers 215, 216, 217 and 218 in this number chart?

Which of these numbers is 12 more than 205?

How do you know?

Extending prompt

Can you find more than 20 possible answers? Convince me you could find all the possible answers by describing the patterns you can see.

CONSOLIDATING THE LEARNING

Four further tasks consolidate the learning from the 'Two numbers' task. The first two tasks are focused on difference. 'A difference of 21' is similar to the original task, whilst '75 apart' requires students to explore a similar idea on a number chart that counts by 10s. The two 'Mr Forgetful' tasks get students to explicitly focus on diagonal patterns, to further support links to additive thinking and number patterns (e.g. counting by 11s, counting by 9s).

A DIFFERENCE OF 21

I am thinking of two numbers on the number chart on page 64.

The difference between my numbers is 21.

One of my numbers has a 3 in it.

What might be the numbers?

75 APART

I am thinking of two numbers on the number chart on page 64.

The difference between my numbers is 75.

One of my numbers has a 6 in it.

What might be the numbers?

Mr Forgetful can't remember how he created this pattern. All he knows is that he started in one of the corners and kept the difference between each number the same.

How might he have created it?

301									
	312								
		323							
			334						
				345					
					356				
						367			
							378		
								389	
									400

How many different diagonal patterns can you create on this chart?

Can you describe the patterns?

Mr Forgetful can't remember how he created this pattern; this time he can only remember some of the numbers in his pattern. All he knows is that he started in one of the corners and kept the difference between each number the same.

How might he have created it?

?									
	110								
		?							
			?						
				440					
					?				
						?			
							?		
								?	
									990

How many different diagonal patterns can you create on this chart?

Can you describe the patterns?

SUGGESTION 6: ESTIMATING NUMBERS ON A NUMBER LINE

MATHEMATICAL FOCUS

A number line is a tool to represent the relative size of a number.

PEDAGOGICAL FOCUS

This suggestion introduces the number line as a way of representing the size of a number. Although the number chart is more useful for young students because it makes the place value structure explicit, the number line is an alternative for representing the relative magnitude of numbers. It is particularly useful for comparing larger numbers.

The focus is getting students to estimate rather than compute using a number line. The number line is not used in this context as a counting tool, nor is it used to support addition and subtraction strategies (although students might choose to use it this way if they revisit Suggestions 4 and 5 after this one). It is to give students opportunities to consider the relative size of numbers and develop their general number sense. Using a number line in this manner is conceptually challenging, as it requires some sense of proportional reasoning. However, it is a natural extension of considering number charts.

Prior to launching the tasks in Suggestion 6, introduce the scaled number line. If students are not familiar with number lines as representations of magnitude (scaled number line), explore/ discuss some examples as a class using smaller numbers. Attempting to count along a number line is not useful when using it to represent magnitude; instead, students need to consider the whole number line and, in particular, the values at either end of the number line (when provided). Questions like the following might be posed using the accompanying number line:

What number might be at Point A? What number might be at Point B? Explain your reasoning.

Prior to launching the 'What numbers might they be?' task, play a 'Guess my number' game with a number line, rather than a number chart (the size of the numbers can be modified to whatever level of challenge is desired). This game can be revisited many times.

The following partially structured number line could be adapted for a 'Guess my number' game appropriate to your students (numbers at either end can be changed to make it more or less challenging). Numbers from an earlier guess should be removed. Questions being posed should focus on the language of 'more' and 'less', for example, *Is your number less than 50?*

WHAT NUMBERS MIGHT THEY BE? ❶

What might be the value of A, B, C and D?

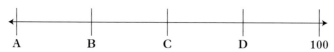

Give more than one set of possibilities.

Explain your reasoning.

Comment

There are many possible sets of answers to the 'What numbers might they be?' task. The main thing for students to notice is that the five points marked on the number line are equally spaced, which means that the difference between consecutive points needs to be consistent for a given set of answers. For example, the difference might be one, leading to one set of answers: A = 96, B = 97, C = 98, D = 99. Alternatively, the difference might be five, leading to another set of answers: A = 80, B = 85, C = 90, D = 95; or 25, leading to another set of answers: A = 0, B = 25, C = 50, D = 75.

Enabling prompt

What number might be at points A and B? Explain your reasoning.

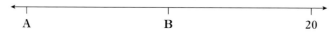

Extending prompt

One of the unknown numbers is 40. What might be the values of A, B, C and D?

CONSOLIDATING THE LEARNING

Four further tasks consolidate the learning. The first is similar to the initial task, except the anchoring number at the far right is 1000, rather than 100. Depending on the cohort/grade-level of students, and how they find the first task, you might choose to use a smaller anchoring number instead (e.g. 50 or 200).

WHAT NUMBERS MIGHT THEY BE? ❷

What might be the values on A, B, C, D?

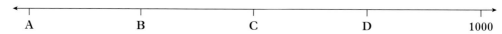

WHAT NUMBERS MIGHT BE A AND B?

What is your best estimate of points A and B on this number line? Explain your reasoning.

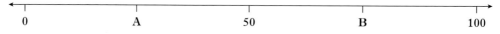

OXFORD UNIVERSITY PRESS

Comment

One way of supporting students to reason proportionally is to encourage them to use half as a benchmark, and then one-quarter (halfway to a half), and three-quarters (halfway between one half and the end). The second consolidating task provides students with an opportunity to explore this idea directly. It differs from the main task in that it only asks students to provide one possible number to represent each letter. Point A equates to the number 25 (it is approximately halfway between 0 and 50), while Point B equates to the number 75 (it is approximately halfway between 50 and 100). Any estimate in the 20s for Point A, and in the 70s for Point B, can be considered sufficiently accurate to be correct. Asking for estimates prompts multiple possible answers that are potentially acceptable and is an interesting point for discussion.

The next three consolidating tasks, 'Missing CD', 'Missing EF' and 'Missing GH' are similar in structure to the previous task, but 50 is not provided as a benchmark.

MISSING CD

What is your best estimate of points C and D on this number line? Explain your reasoning.

MISSING EF

What is your best estimate of points E and F on this number line? Explain your reasoning.

MISSING GH

What is your best estimate of points G and H on this number line? Explain your reasoning.

CHAPTER 6

INFORMAL LENGTH MEASURING

OVERVIEW

This sequence is suitable for students from Foundation to Year 2. The first three suggestions are suitable for all years, including Foundation, while the latter suggestions are suited for Years 1 and 2.

This sequence lays the foundation for formal units. Even though formal units are not strictly part of the curriculum at Year 2, many students in Year 2 are ready and keen to use formal units (cm and m). The latter suggestions in this sequence can be adapted to allow students to answer the questions using formal units.

This sequence provides important preliminary experiences for the 'Informal approaches to perimeter and area' sequence and the 'Volume' sequence.

The following is a summary of the suggestions in this sequence.

	MATHEMATICAL FOCUS
Direct comparisons	Lengths can be compared by placing one object against another.
Indirect comparisons with lines	Lengths can be compared by using a third object.
Indirect comparisons with different shapes	A third object is used to compare one dimension of an object against another.
Using informal units iteratively	The same object can be used repeatedly to compare lengths.
Using informal units to compare different objects	When comparing different informal units, the unit in each case must be constant.

RATIONALE

The focus of this sequence is on informal measurement prior to formal units. This includes visual and direct comparisons, indirect comparisons, and using informal units iteratively to compare. The intention is to develop an intuitive sense of length, and it is helpful to prompt students to estimate before measuring. Encourage students to use specific language associated with measurement, and provide them with ample opportunities to reason when making comparisons.

It is important to note that students do not need to be proficient with counting prior to introducing this sequence, as learning to measure is dependent on experiences related to measurement concepts.

Since the tasks are experiential, there is likely to be no need for enabling prompts. After completing each task, students should be encouraged to form generalisations related to measuring.

OXFORD UNIVERSITY PRESS

LANGUAGE

long, longer, longest, short, shorter, shortest, high, height, wide, width, deep, depth, horizontal, vertical, circumference

LAUNCHING THE TASKS

You might lead discussions intended to connect length measuring with students' experience. You could ask questions, such as, *Have you seen anyone measuring how long something is?*

In the early years, teachers have found that stories that connect with comparing and measuring length are useful. Examples of books that are useful are *The King's Foot* and *The Long Red Scarf.*

ASSESSMENT

Suggested pre-test and post-test assessment task: Suggestion 2: 'Comparing lines that are not straight (1) and (2)'.

The following are some specific statements to inform assessment.
Students are learning about length when they can:

- estimate and compare objects using direct comparisons based on their length
- estimate, compare and order objects using indirect comparisons based on their length
- estimate, compare and order lengths using uniform informal units
- solve problems related to comparing and ordering lengths using uniform informal units.

SUGGESTION 1: DIRECT COMPARISONS

MATHEMATICAL FOCUS

Lengths can be compared by placing one object against another.

PEDAGOGICAL FOCUS

This task is intended to provide an experience of comparison, and requires both direct and physical comparison. The students will have hand spans that are quite similar, so accuracy is important. All of the comparisons and diagrams are illustrative, with the intention being that you create your own worksheets.

When launching the first task, encourage students to stretch their hands as wide as they can to determine their hand span.

COMPARING HAND SPANS

Who has a hand span the same as yours? Whose is shorter?
What do you have to think about when comparing hand spans?

hand span

Comment

Take any opportunity to extend the discussion; for example, discuss what
the attribute is (i.e. length) and strategies for comparing lengths accurately.

CONSOLIDATING THE LEARNING

There are two further tasks providing experiences for estimation and direct comparisons using
students' hands and feet, as well as external objects.

FIND SOMETHING

Find something that is longer than your hand span, but shorter than your foot.

STREAMER

Tear a streamer that is longer than the blue one but shorter than the red one.

SUGGESTION 2: INDIRECT COMPARISONS WITH LINES

MATHEMATICAL FOCUS

Lengths can be compared by using a third object.

PEDAGOGICAL FOCUS

The term 'indirect' is for those situations in which a third instrument (e.g. a long piece of string
or streamer) is needed to compare lengths.

For some students, the task is an optical illusion. The purpose of the illusion is to emphasise the
importance not only of estimating but also of measuring for accurate comparisons, and the need
for a third object(s) for the comparison. Encourage students to find different ways of doing this.

Note that the lines are only illustrative. Create your own models for the students. Some tasks
involve ordering more than one length.

You will need to work out the optimal ways to pose the tasks, such as using an A3 size version,
or large-scale posters.

OXFORD UNIVERSITY PRESS

HORIZONTAL OR VERTICAL LINES

Guess which is longer: the horizontal or vertical line?

How could you work out which line is longer?

Comment

The horizontal line shown here is longer, even though it does not look like it.

CONSOLIDATING THE LEARNING

There are further tasks that provide experiences for indirect comparisons. The first two tasks require students to indirectly compare and order straight lines. The next two tasks extend this thinking to curved lines, and the last task makes comparisons between the length around an object with straight lines.

ARRANGE THESE LINES IN ORDER OF LENGTH

Guess which line is the longest. Which is the shortest? Check your guess.

A

B

C

D

E

F

Comment

Provide students with a copy of **BLM 9: Arranging lines in order of length**.

Guess which two blue lines are the same length.

BLM 9

Oxford
OWL

Check your guess.

> —————— A ——————<
< —————— B ——————>

< —————— C ——————>
< —————— D ——————>

COMPARING LINES THAT ARE NOT STRAIGHT ❶

Draw a line longer than the blue one but shorter than the red one.

COMPARING LINES THAT ARE NOT STRAIGHT ❷

Arrange these lines in order of length.

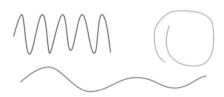

FOOT OR PACE

Work out which is longer: the distance around your foot or the length of one pace (step).

SUGGESTION 3: INDIRECT COMPARISONS WITH DIFFERENT SHAPES

MATHEMATICAL FOCUS

A third object is used to compare one dimension of an object against another.

PEDAGOGICAL FOCUS

As in the previous suggestion, a third tool (e.g. a long piece of string or streamer) is needed to compare the lengths. Later, rulers and tape measures can be used, but at this stage it is better to establish the basic concept first. Note that the above task is the third task in this suggestion.

It is important for students to estimate and discuss measurement strategies prior to actual comparisons. The key idea is that students are comparing different dimensions of the same object. Use real containers that are familiar to students, rather than the pictures. Encourage students to generalise their strategies.

COMPARING CIRCUMFERENCE AND HEIGHT ❶

Which do you think is longer: the circumference of this container (at the wide part) or its height?

Measure the height and circumference to decide which is longer (do this in two different ways).

Comment

As with an earlier task, it is quite hard to estimate which is longer. This emphasises the importance of checking the accuracy of estimates.

CONSOLIDATING THE LEARNING

There are two tasks providing experiences with estimation, comparisons and ordering that come before 'Comparing circumference and height (2)'. The tasks with three common containers involve first imagining and estimating, then comparing the actual containers.

IMAGINING AND COMPARING THE HEIGHT OF CONTAINERS

Imagine which of these is the tallest. Which do you think is the shortest?

Now measure them to check your estimates.

IMAGINING THE CIRCUMFERENCE

Imagine which of these has the longest circumference (at the widest spot).

Imagine which of these has the shortest circumference (at the widest spot).

Now measure the circumferences to check your estimates.

Which do you think is longer: the circumference of this container (at the wide part) or its height?

Measure the height and circumference to decide which is longer (do this in two different ways).

SUGGESTION 4: USING INFORMAL UNITS ITERATIVELY

MATHEMATICAL FOCUS

The same object can be used repeatedly to compare lengths.

PEDAGOGICAL FOCUS

Using a (uniform) unit iteratively is the next step in learning about length.

There are many possibilities for the 'Michael and Monica' problem, and it might help to make a list of the rules generated from earlier experiences (for example, 'do not leave gaps' or 'measure in the shortest straight line'). Matchsticks, or anything linear, such as craft sticks, are an ideal classroom resource for this suggestion since they prompt 'length' (whereas the dimensions of paper clips, Unifix, etc. are ambiguous).

MICHAEL AND MONICA

Michael measured his desk and said it was 20 craft sticks wide.

Monica measured the desk and said it was 19 craft sticks wide.

How could this happen?

Give 5 different possibilities.

Comment

Take opportunities to extend the discussion. For example, one of the students might suggest that Michael left gaps. Have a class discussion to establish how we know Monica was the person who left gaps.

CONSOLIDATING THE LEARNING

There is one task before the 'Michael and Monica' task that requires using similar iterative units: '50 matchsticks'. Note that the final task ('Half your height') is more difficult, but students should be ready for this after working on earlier experiences.

50 MATCHSTICKS

Write your name using 50 matchsticks.

WHO LEFT GAPS?

Either Michael or Monica left gaps between the craft sticks. Who left the gaps?

HALF YOUR HEIGHT

How might you work out how old someone would be who is half your height?

SUGGESTION 5: USING INFORMAL UNITS TO COMPARE DIFFERENT OBJECTS

MATHEMATICAL FOCUS

When comparing different informal units, the unit in each case must be constant.

PEDAGOGICAL CONSIDERATIONS

The tasks emphasise student decision-making for choosing a unit that is constant to compare different objects. It is important for students to estimate and discuss measurement strategies prior to actual comparisons. Use real objects from the classroom and invite students to imagine the answers in each case before trying them out. If students are ready, you can adapt the following tasks to allow students to answer the questions using formal units.

When launching this suggestion, you could revisit the list of the rules generated from earlier experiences (for example, 'do not leave gaps' or 'measure in the shortest straight line').

FEET AND DESKS

Guess how many feet, placed heel to toe, are the same length as two desks.
Now do it.

Comment

As with earlier tasks (see Suggestions 2 and 3) it is quite hard to estimate which is longer. This emphasises the importance of checking the accuracy of estimates.

CONSOLIDATING THE LEARNING

There are three further tasks that extend the learning from the above task. Note that the final task is more complex than the previous ones.

ARM SPANS AND FEET

Guess which would be longer, 2 of your arm spans (tip to tip) or 10 of your feet placed end to end. Now measure which is longer.

HIGHLIGHTER PENS OR SHEETS OF PAPER

Guess which is longer, 7 highlighter pens (tip to tip) or 3 A4 sheets of paper. Now measure which is longer.

IF WE MEASURED IN ACTUAL FEET

Who would be tallest if measured by the number of their feet lengths?
Who would be shortest if measured by the number of their feet lengths?

CHAPTER 7

TIME

OVERVIEW

This sequence is suitable for students from Foundation to Year 2. The first three suggestions are suitable for all years and can be modified to ensure the language, length of time intervals and degree of accuracy required in telling the time on a clock are appropriate for students. The latter suggestions are suited to Years 1 and 2. The focus of Suggestion 2 is on informal measurement of time.

Being able to read time on a clock, or the date on a calendar, is reliant on students' knowledge of numbers (particularly for Suggestions 3 and 4). If students do not possess the prerequisite number knowledge, this could be intentionally developed alongside specific suggestions in this sequence by revisiting tasks from number sequences.

The following is a summary of the suggestions.

	MATHEMATICAL FOCUS
The language of time	The language of time includes particular terms, as well as everyday words and phrases.
Duration	Duration of events can be compared and ordered using everyday language.
Analogue clocks	Clocks are used to tell the time and to measure how long (duration) an action or event takes.
Calendars	Calendars show particular periods of time. They can be used to identify dates and events, and to measure passages of time.

RATIONALE

The focus in this sequence is on developing four big ideas of time:

- an awareness of time
- ordering or sequencing events in time
- duration of events
- the measurement of time (including using units of time).

Developing the *language of time* is fundamental to each of these ideas. Importantly, the development of these big ideas is more interdependent than sequential. For example, growth in students' capacities to measure time using a clock is simultaneously reliant on growth in their sense of time, their understanding of duration and an expanding familiarity with, and ease in using, a range of units of time. For this reason, students could benefit from revisiting some of the suggested tasks throughout Foundation to Year 2.

Many of the tasks use examples that should be possible at school. However, they are illustrative, and it is assumed you will adapt them to suit your own students' contexts (while preserving the nature of the challenge). Due to the tasks having both low floors and high ceilings, enabling and extending prompts are provided only for the first task in this sequence.

LANGUAGE

Descriptive words: day/daytime, night/night-time, early, late, morning, noon, afternoon, midnight, midday, every day, today, tomorrow, yesterday, sleeps, days of the week, seasons
Comparative words: earlier, later, longer, shorter, faster, slower, before, after, next
These lists are not exhaustive; you can add to them as necessary.

LAUNCHING THE TASKS

Quality stories that connect with telling the time and measuring time are helpful. Examples of books are *Diary of a Wombat* by Jackie French (events in the day and days of the week), *The Very Hungry Caterpillar* by Eric Carle (days of the week), *The Grouchy Ladybug* by Eric Carle (telling time on the hour), and *Clocks and More Clocks* by Pat Hutchins (measuring time and duration of time). Playing popular children's games (e.g. 'What's the time Mr Wolf?' and 'Time bingo') and singing songs (e.g. 'Rock around the clock') will help students to consolidate important time language. Display a hard copy calendar, analogue clock and digital clock in your classroom. Refer to them throughout the day to familiarise students with their features and contextualise their usefulness in everyday life.

ASSESSMENT

The time sequence is comprised of various aspects; it is therefore important to consider the concept you intend to assess. For example, duration (suggested pre-test and post-test assessment task: Suggestion 3: 'Analogue clocks').

The following are some specific statements to inform assessment.
Students are learning about time when they can:
- use everyday language of time to name events
- sequence routine daily events in time
- sequence familiar events of the week and year
- connect familiar events to specific times of the day, week and year
- recognise events that have a duration of up to/more than one minute and less than one hour
- estimate how long things take
- compare and order the duration of two events using everyday time language
- describe how long things take using informal and formal units of time
- read an analogue clock to the hour, half-hour and quarter-hour
- describe and explain time according to the positioning of the hands on a clock
- calculate, compare and explain the duration (or difference) of two times
- use a calendar to find a particular date or day
- determine the number of days in each month
- measure and justify the duration of an event using days, weeks and months.

MATHEMATICAL FOCUS

The language of time includes particular terms, as well as everyday words and phrases.

PEDAGOGICAL FOCUS

The focus in this suggestion is on the language of time, grouping terms together and then sequencing events in time. Start with students' own suggestions for events in time based on their everyday routines. Supplement this list with terms from the 'Language' section at the start of this sequence. The list of terms is not exhaustive; it is recommended that you add or remove terms as appropriate.

TIME WORDS

What different words do you know for parts of the day?
Mark on a timeline the range of times that these words can apply to.

midnight   midday   midnight

Comment

Students' awareness or sense of time develops as they:

- use words that describe points in time, events and routines
- arrange connected events in the usual sequence that they occur
- compare the duration of everyday events using everyday language of time
- use units of time.

Recording the sequence of time words on a strip of paper (between 50–100 cm in length) will allow the two 'midnight' ends to overlap. Once the ends are taped together the strip will form a loop, reinforcing the idea that many routine events occur each day. This circular representation might also stimulate discussion about how a day is measured from midnight to midnight.

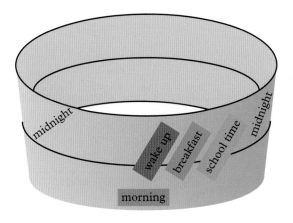

Enabling prompt

Use pictures to accompany unfamiliar time words.

Mark 'dinner time' on this timeline.

midnight   midday   midnight

Extending prompt

Find some words for parts of the day in a language other than English.

Find the origins of the meaning of the words for the days of the week.

BLM 10
BLM 11

Oxford
OWL

CONSOLIDATING THE LEARNING

There are further tasks that consolidate the learning from the 'Time words' task, which use different time words. It is expected that most students can find multiple ways to group the words and possibly even sequence them according to their routines. For each task, students should be encouraged to explain and justify their groupings and sequences.

Give students a copy of **BLM 10: Time words (1)** and **BLM 11: Time words (2)**.

SORT, DESCRIBE AND ORDER TIME WORDS ❶

Sort the following time words into groups. Give a name to each group. Arrange the words in each group in the order in which they occur.

Choose one of the cards and draw a picture to show what is happening at that time.

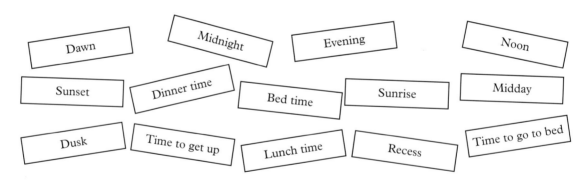

Dawn   Midnight   Evening   Noon

Sunset   Dinner time   Bed time   Sunrise   Midday

Dusk   Time to get up   Lunch time   Recess   Time to go to bed

PARTS OF THE YEAR

What are some different words for parts of the year?

Mark on a timeline the range of times that those words can apply to.

January   December

SORT, DESCRIBE AND ORDER TIME WORDS ❷

Sort the following time words into groups. Give a name to each group. Arrange the words in each group in order.

Choose one of the cards and draw a picture to show what is happening at that time.

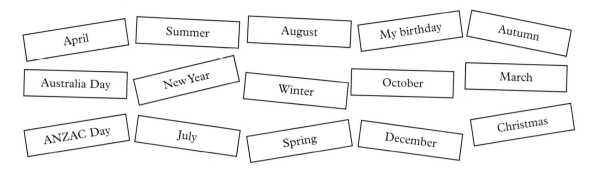

TIME SENTENCE

Write a sentence that makes sense that has as many time words as possible.

NAMES

What are some names of people that are also time words?

SUGGESTION 2: DURATION

MATHEMATICAL FOCUS

Duration of events can be compared and ordered using everyday language.

PEDAGOGICAL FOCUS

These tasks have multiple possible answers that students should physically experience; for example, how long (duration) it takes to walk around the basketball court. It is assumed you will use a digital stopwatch or minute sandglass timer (virtual or physical version) as a way of measuring minutes. The tasks in this suggestion require students to estimate and compare, with the focus more about the justification of choices than being right or wrong.

Give students a copy of **BLM 12: Estimate then measure**. This BLM can be modified as needed to ensure the relevance to the students.

Prior to launching this suggestion, you could ask students to stand up, close their eyes and sit down when they think one minute has elapsed. During the launch, select a student to physically eat an apple (or similar food/activity that corresponds to the task presented) while a timer is visible for all students to see.

BLM 12

Oxford
OWL

LONGER THAN A MINUTE

What is something that takes you longer than 1 minute but shorter than the time it takes you to eat an apple?

Comment

The intention is that students perform all activities at their normal pace unless challenged to do them quickly. The time taken to do activities will vary for different students. Activities could include reading a page or two of a storybook or clearing away mathematics equipment.

CONSOLIDATING THE LEARNING

Further tasks are proposed that are intended to consolidate students' estimation of how long things take, and to compare and order the duration of events using different time measures. Students should be encouraged to use time words and justify their answers for all tasks.

LESS THAN 1 HOUR

What is something that takes longer than recess but less than 1 hour?

HOW LONG DO THESE THINGS TAKE?

Estimate then measure how long each of these things take:
- Stand on one foot without falling over
- Write your full name as neatly as possible
- Run the length of the basketball court
- Count to 100 by 10s, saying all the numbers clearly
- Count to 100 by 1s, saying all the numbers clearly
- The amount of time before someone walks into the room.

10 TIMES IN A MINUTE

What is something that you can do 10 times in a minute going as fast as you can?

Estimate then measure how many times you can do each of the following in one minute:
- Times you can say "she sells sea shells" correctly
- Times you can write your name neatly
- Times you can do a star jump
- Times you can clap
- Times you can throw a ball to a partner and catch it when it is thrown back.

WHAT DO PEOPLE MEAN WHEN THEY SAY ...?

"We need to leave in 5 minutes." "You have to go to bed soon." "We can do that later." "Just a second." "Wait a minute."

MATHEMATICAL FOCUS

Clocks are used to tell the time and to measure how long (duration) an action or event takes.

PEDAGOGICAL FOCUS

This suggestion focuses on how the hour hand helps us to tell the time, and supports students to use language such as 'hour hand' and 'minute hand' (not 'little' and 'big' hands). The tasks also further develop the idea of duration by asking students to calculate the difference between two times. Discuss with students why the time difference questions all have two answers.

CONSOLIDATING THE LEARNING

The following tasks are intended to consolidate students' capacities to tell time using an analogue clock and to use clocks to measure time (duration) beyond time on the hour. Students should be given opportunities to construct their own clocks and to physically manipulate the arms of a mechanical clock.

Prior to launching the suggestion, you could ask students to draw a clock. Begin by ensuring all clocks in the room are hidden. Next, discuss students' clocks and the times they may have drawn on them. After students have discussed their clocks, you could also bring out a variety of clocks to discuss the differences and similarities between them.

CLOCKS

Draw in the hour hand to show your favourite time of the day.

What time does your clock show?

Now draw what it would look like 1 hour later.

Give students a copy of **BLM 13: Analogue clocks**.

BLM 13

Oxford OWL

Comment

Encourage students to go beyond using times on the hour ('o'clock times'). Students should justify how they know that the second clock shows time that is one hour later than the first. What has changed?

HOW LONG IS IT BETWEEN THESE TIMES? ❶

How long might it be between these two times?

Find as many different answers as you can.

HOW LONG IS IT BETWEEN THESE TIMES? 2

How long might it be between these two times?
Find as many different answers as you can.

GONE TO LUNCH

I went to my favourite toy store to buy a birthday present. I arrived at the store at 12.30 pm, but the door was locked, and this sign was in the window.

What time will the shopkeeper return from lunch?

How long do I have to wait?

Justify your answer.

Oxford
OWL

ONE-HANDED CLOCKS 1–3

Give students a copy of **BLM 14: One-handed clocks**.

I have an old clock that only has the hour hand, but I still think I can use it to tell the time.

How long might it be between these two times? (There are two possible answers in each case.)

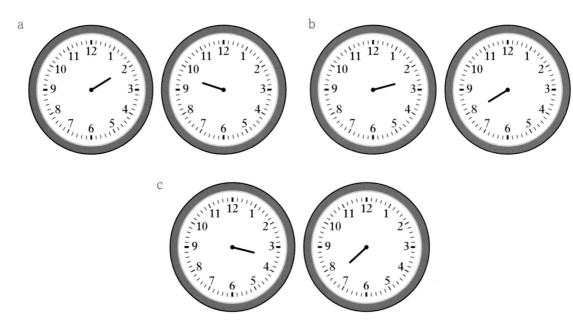

SUGGESTION 4: CALENDARS

MATHEMATICAL FOCUS

Calendars show particular periods of time. They can be used to identify dates and events, and to measure passages of time.

PEDAGOGICAL FOCUS

It is assumed that students will have access to this year's calendar. Most tasks have multiple possible answers that will vary depending on the year and/or month, so students will have something to explain to the class. Students should explore what differences occur when a different year or month is the focus, and why these differences occur.

Prior to launching the suggestion, ensure students are familiar with a hard copy calendar – you may wish to provide them with a copy of **BLM 15: Calendar**. You could refer to the calendar in previous weeks and days leading up to this suggestion, asking students to locate 'today's date' and asking students what they notice and what they wonder about the calendar. For example, students could note that the week starts on a Sunday, but the first day of every month is not always a Sunday (*I wonder why!*). To extend their familiarity with the structure of a calendar and the order of the months, you could cut up an old calendar and ask students to reconstruct it. When launching the suggestion, clarify if students understand the information that each aspect of the calendar is conveying. For instance, that the letters 'S', 'M', 'T' etc. stand for the days of the week; that they know which 'S' stands for 'Saturday' and which stands for 'Sunday'.

BLM 15

Oxford
OWL

5 THURSDAYS AND 5 FRIDAYS

Which months have 5 Thursdays and 5 Fridays? What do these months have in common?

CONSOLIDATING THE LEARNING

The following tasks are similar in that they ask students to determine dates on a calendar but also require them to determine duration of time. In each case, there are multiple answers, and students should justify their responses. Answers will vary according to the year, so it is important that teachers do these tasks themselves before giving them to students, to ensure they are still applicable. For the last task, students could use a calculator to convert units of time (days to hours). Prior to teaching, check that the dates/days chosen are suitable for the current year's calendar (e.g. they have multiple possible answers).

TUESDAY THE 19TH

Tuesday is the 19th. What month could it be?

How many possible answers are there?

What do those months have in common?

FRIDAY THE 24TH

I am thinking of two dates.

The dates are 25 days apart.

One of the dates is Friday 23rd.

What might be the dates?

How many answers are possible?

Why are there so many different possible answers?

HOW MANY DAYS?

How many days is it until your next birthday?

How can you work that out without counting every day?

WHAT DATE?

What date might I be?

My month has 30 days.

I am the 3rd Saturday in my month.

IN 2030

Work out how many days it is from June 29th to September 7th in 2030 without using a calendar.

How many hours is that?

CHAPTER 8

INFORMAL APPROACHES TO PERIMETER AND AREA

OVERVIEW

This sequence is suitable for students in Years 1 and 2. The first three suggestions are suitable for both years, while Suggestions 4 and 5 are more suitable for Year 2, and would also be suitable for Year 3.

This sequence lays the foundations for formal units and builds on length measurement. It is suggested that students complete the 'Informal length measuring' sequence prior to starting this sequence. The goal is to extend students' intuitive understanding of perimeter and area using informal approaches. There are advantages in students learning perimeter and area at the same time, so they can compare and contrast the concepts.

This sequence includes ideal preliminary experiences to the 'Volume' sequence.

The following is a summary of the suggestions in this sequence.

	MATHEMATICAL FOCUS
Perimeter	Perimeter is the length of a shape's boundary. It can be measured using uniform informal units.
Area	Area is the measure of how many square units are needed to cover a surface.
Contrasting perimeter and area	Perimeter and area are measured differently.
Same area, different perimeters	Shapes with the same area do not necessarily have the same perimeter.
Same perimeter, different areas	Shapes that have the same perimeter might have different areas.

RATIONALE

The focus of this sequence is on informal measurement prior to formal units. This includes exploring the relationship between perimeter and area and comparing measurements. Through experiences of measuring perimeter (as an extension of length measurement) and area of shapes, it is intended that students build the idea that both have different measures and use different units. Common misconceptions include a belief that for any one area there is only one perimeter. The following suggestions help to explore this idea in the early years so that these misconceptions do not develop.

Due to the tasks having both low floors and high ceilings, there are no enabling prompts suggested, and few extending prompts. For students who find these tasks straightforward, increase the size of the shapes. Also, encourage them to form generalisations about the relationship between area and perimeter. These generalisations include knowing when we might want to measure perimeter and/or area, what shapes can be measured easily, efficient ways of recording, and ways of calculating perimeter and area from diagrams.

LANGUAGE

perimeter, area, around, distance, surface, amount covered, dimensions, circumference, rectangle, square, oblong, measure, length, width, rows, columns, array, unit, overlap, gap

ASSESSMENT

Suggested pre- and post-test assessment task: Suggestion 3: 'A rectangle of 16 squares'.

The following are some specific statements to inform assessment.
Students are learning about perimeter and area when they can:

- gesture to identify the perimeter as the distance around
- construct and describe the perimeter of shapes
- estimate and compare perimeters using length
- create, explain and systematically record several possibilities for a given perimeter
- gesture to identify the area of a rectangular shape as the surface covered
- construct and describe the area of rectangular shapes
- estimate and compare areas using square units
- create, explain and systematically record several possibilities for a given area
- construct and describe shapes with the same area but different perimeter
- construct and describe shapes with the same perimeter but different area
- explain how and why perimeter and area are measured differently.

SUGGESTION 1: PERIMETER

MATHEMATICAL FOCUS

Perimeter is the length of a shape's boundary. It can be measured using uniform informal units.

PEDAGOGICAL FOCUS

Students use informal units to measure length. Matchsticks, craft sticks or similar are ideal. Encourage students to record their answers. Squared, grid or dot paper might help for recording.

The tasks have multiple possible answers that the students can create using physical models if needed. During the launch of the suggestion, students could walk around the outside of a basketball court, or you could find or create a relevant story.

HOW FAR AROUND A SHAPE? 1

12 sticks have been used to make this rectangle.
Make a rectangle using 16 sticks.
Do this in a few different ways.
Record your answers.

Comment

A variety of rectangles can be made using whole stick lengths. Rectangles could be made in the shape of oblongs, such as seven sticks long and one wide, six long and two wide, five long and three wide, or as a square of four long and four wide. This possibility might allow you to discuss the relationship between squares and oblongs, since both belong to the family of rectangles.

CONSOLIDATING THE LEARNING

There are further tasks that explore the idea of perimeter as the distance around a shape. The first two consolidate the learning from the first task. In these tasks, encourage students to use the sticks as accurately as possible (no big gaps or overlaps and laid straight). The other two tasks extend that learning by moving to composite shapes.

HOW FAR AROUND A SHAPE? 2

8 sticks have been used to make this triangle.
Make a triangle using 12 sticks.
Record your answers.
Do this in a few different ways.

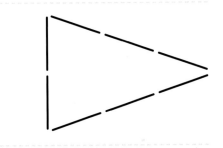

HOW FAR AROUND A SHAPE? 3

10 sticks have been used to make this trapezium.
Make a trapezium using 12 sticks.
Record your answers.
Do this in a few different ways.

MAKE AN L 1

This "L" shape has a perimeter of 12 units (or sticks).
Make a shape like an "L" with a perimeter of 16 units.
Record your answers.
Do this in a few different ways.

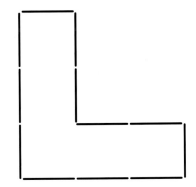

This "L" shape has a perimeter of 12 units.

Make a shape like an "L" with a perimeter of 20 units.

Record your answers.

Do this in a few different ways.

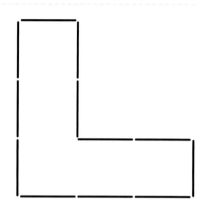

SUGGESTION 2: AREA

MATHEMATICAL FOCUS

Area is the measure of how many square units are needed to cover a surface.

PEDAGOGICAL FOCUS

The tasks in this suggestion emphasise student decision-making. Flat square counters (or squares cut up) are ideal. Squared, grid or dot paper can be used for recording.

Some students might count and compare to explore possibilities. For students who find all possible solutions, prompt them to make generalisations and increase the number of square counters.

The tasks in this suggestion have multiple possible answers that the students can create using physical models if needed. Use flat counters rather than any with a raised surface (height) to help prevent misconceptions or confusion arising later when students are working with volume (three dimensions).

12 SQUARES

8 square counters have been used to make this rectangle. Make a rectangle using 12 square counters. Do this in a few different ways. Record your answers.

Comment

Assuming square units are used, the rectangles can be six counters long and two wide (you can discuss whether two long and six wide is the same or different), four long and three wide, twelve long and one wide. Some students might seek to use half squares – if they do, make sure they do this accurately (it is quite complex).

CONSOLIDATING THE LEARNING

Further tasks that explore the idea that area is the amount of surface covered and can be measured in squares are as follows. The first one is to consolidate the learning from the above task. The next two extend the learning to composite shapes.

20 SQUARES

8 square counters have been used to make this rectangle.

Make a rectangle using 20 square counters.

Record your answer. Do this in a few different ways.

AN L WITH 20 SQUARES

Make a shape like an "L" using 20 square counters. Record your answer.

Do this in a few different ways.

AN L WITH 24 SQUARES

Make a shape like an "L" using 24 square counters. Record your answer.

Do this in a few different ways.

SUGGESTION 3: CONTRASTING PERIMETER AND AREA

MATHEMATICAL FOCUS

Perimeter and area are measured differently.

PEDAGOGICAL FOCUS

The intention is to move students beyond having them make the shapes to recording their thinking. Squared, grid or dot paper will help with recording. Encourage students to do these tasks more than one way, find all possible solutions, make generalisations and explore using a larger number of squares.

When launching this suggestion, it is important to discuss the fact that a single square counter has an area of one unit, but the perimeter is four units. Students could work out the perimeter and area of two squares together. Consider having some students standing inside the basketball court and others standing around the edge.

A RECTANGLE OF 16 SQUARES

8 square counters have been used to make this rectangle.

The perimeter is 12 units.

Make a rectangle using 16 square counters. Draw your rectangle.

What is the perimeter of your rectangle?

Make some different rectangles.

Comment

For a rectangle that is 16 units long and one wide, the perimeter is 34 units. For eight long and two wide, the perimeter is 20 units; for four long and four wide, the perimeter is 16 units.

CONSOLIDATING THE LEARNING

There are further tasks proposed in this suggestion that explore the idea that perimeter and area are measured differently and use different units. The first task consolidates the learning from making a new rectangle. The next two tasks are further examples of the same thinking. The final two tasks extend the thinking beyond the square counters or pictures that are counted one by one, to thinking about dimensions.

A RECTANGLE OF 24 SQUARES

8 square counters have been used to make this rectangle.

The perimeter is 12 units.

Make a rectangle using 24 square counters.

Draw your rectangle.

No gaps or overlaps are allowed.

What is the perimeter of your rectangle?

Make some different rectangles.

PERIMETER OF 16 UNITS

Use square counters to make a rectangle (no holes in the centre) with a perimeter of 16 units. Draw your rectangle.

Do this in a few different ways.

PERIMETER OF 18 UNITS

Use square counters to make a rectangle (no holes the centre) with a perimeter of 18 units. Draw your rectangle.

Do this in a few different ways.

OUT OF 100 SQUARES

Without actually making it, what might be the perimeter of a rectangle made out of 100 squares? Explain your reasoning.

Do this in a few different ways.

OUT OF 80 SQUARES

Without actually making it, what might be the perimeter of a rectangle made out of 80 squares? Explain your reasoning.

Do this in a few different ways.

MATHEMATICAL FOCUS

Shapes with the same area do not always have the same perimeter.

PEDAGOGICAL FOCUS

These tasks provide a different perspective on perimeter and area. The idea is for students to imagine what might happen if they cut the shape. Students who need it can use squared paper. Squared counters will also work for the tasks.

MAKING A NEW RECTANGLE ①

How many squares are needed to make this rectangle?

What is the perimeter?

Imagine that you cut it once and rearrange it to make a new rectangle using all the squares.

How many squares are in the new rectangle?

What is the new perimeter?

Do this in a few different ways.

Comment

The rectangle is made from 24 squares and has a perimeter of 20 units. There are two solutions to this task. It is possible to cut it vertically, making a rectangle 12 squares long and two wide with the same area, but with a perimeter of 28 units. (Students might notice it is also possible to cut this rectangle a second time, and create a rectangle 24 long and one wide.). It is also possible to cut the original rectangle horizontally, making a rectangle eight wide and three long with a perimeter of 22 units.

CONSOLIDATING THE LEARNING

There are further tasks proposed that explore the idea that perimeter is measured differently and uses different units. The first consolidates the learning from the above task, and the following tasks are posed without a diagram, encouraging students to imagine the rectangles.

MAKING A NEW RECTANGLE ❷

How many squares are needed to make this rectangle? What is the perimeter?

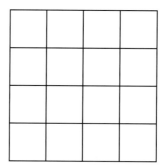

Imagine that you cut it once and rearrange it to make a new rectangle using all the squares. How many squares are in the new shape? What is the new perimeter?

KEEPING THE AREA THE SAME ❶

Draw a rectangle on squared paper that is 10 squares long × 4 squares wide.

Imagine that you cut the rectangle once and reassemble the parts into another rectangle so that the area is the same.

What is the new perimeter?

Do this in a few different ways.

KEEPING THE AREA THE SAME ❷

You have a rectangle on squared paper that is 8 squares long × 4 squares wide.

Imagine that you cut the rectangle once and reassemble the parts into another shape so that the area is the same.

What is the new perimeter?

Do this in a few different ways.

SUGGESTION 5: SAME PERIMETER, DIFFERENT AREAS

MATHEMATICAL FOCUS

Shapes that have the same perimeter might have different areas.

PEDAGOGICAL FOCUS

The focus of this suggestion is on polygons with the same perimeter but with a different area. Students could use squared paper or flat square counters. When launching this suggestion, ensure students are aware that the new polygons they create do not necessarily need to be rectangles. Remind students that the side of each square, for this and subsequent tasks, is 1 unit in length.

We have a rectangle made of 12 squares, arranged 6 × 2.

We want to remove 2 squares but keep the perimeter the same as it is now.

What are some possibilities?

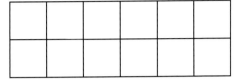

Comment

The perimeter stays the same if any two non-adjacent corners are removed. The perimeter also stays the same if two adjacent squares, one of which is a corner, are removed (again, excluding the two adjacent corners).

CONSOLIDATING THE LEARNING

There are further tasks proposed that consolidate the learning from the above task. These last tasks are more complex than the previous ones.

KEEPING THE PERIMETER THE SAME **2**

We have a rectangle made of squares arranged 4 × 2.

We can remove 2 squares, but we want to keep the perimeter the same as it is now.

What are some possibilities?

KEEPING THE PERIMETER THE SAME **3**

We have a rectangle made of 18 squares arranged 6 × 3. The goal is to remove 2 squares but keep the perimeter the same. What are some possibilities?

CHAPTER 9

VOLUME

OVERVIEW

This sequence is intended to be close to two weeks of mathematics lessons and suitable for students in Years 1 and 2. The first three suggestions are suitable for both years, while Suggestions 4 and 5 are suited to Year 2.

The goal of this sequence is to offer students experiences of volume of rectangular prisms using informal approaches. This sequence lays the foundation for formal units. It is suggested that students complete the 'Informal length measuring', 'Informal approaches to perimeter and area' and 'Objects' sequences first.

The following is a summary of the suggestions in this sequence.

	MATHEMATICAL FOCUS
Investigating the volume of prisms	The volume of a prism can be found by determining the number of cubes needed to build it.
Creating prisms from a description	Rectangular prisms can be described by length, width and height.
Creating different shaped prisms	Prisms with the same volume can have different dimensions.
Imagining what the prism might look like	The dimensions of a prism can be used to create a mental image of what it might look like.
Investigating 'stepped' prisms	There are different ways to determine how many cubes are needed to build 'stepped' prisms.

RATIONALE

The three-dimensional concept of volume refers to the amount of space taken up by the solid object. As volume is an attribute of measurement, it is important to note whether students are recognising the quantity (e.g. 12) and the unit (e.g. cubes).

The concept of volume develops in the same order as other measurement attributes (comparing; informal measuring; formalising measurement; solving problems involving measuring). However, volume is conceptually different from other dimensional attributes, such as length and area, for three reasons. First, it adds a third dimension and consequently extends students' idea of spatial structuring. Second, it uses different units of measurement, that is, cubes. Third, there are some skills and concepts that are developed by considering volume. These include interpreting and giving descriptions of rectangular prisms, interpreting, imagining and creating two-dimensional representations of rectangular prisms, and using and interpreting prisms given the dimensions.

Developing an initial understanding of volume as the amount of space taken up by an object is supported through several experiences, including stacking and packing activities, playing with blocks (wooden cubes, multilink, LEGO), as well as building structures, using the same blocks to make different structures, and describing and comparing them.

The focus of this sequence is on determining the volume of rectangular prisms by estimating, constructing and interpreting representations, and for students to recognise that prisms with the same volume can have different dimensions. Use of comparison language when students describe the size of their constructions and examine which construction takes up the most/least or same amount of space will assist students' developing understanding of volume. The benefits to students doing the tasks in this sequence go beyond developing an understanding of volume. For example, as many cubes are not directly visible in the various prisms, students should be encouraged to visualise the objects in their head. There are many opportunities for students to count the cubes that comprise the prisms by using efficient counting strategies, such as skip-counting, which supports early multiplicative thinking.

LANGUAGE

prism, cubes, rectangular shape, length, width, height, depth, space, measure, volume, and *descriptive language* such as long, wide, deep, high, tall, narrow, more, less, same, layers

Note: language associated with the properties of prisms explored in the 'Objects' sequence can be reinforced in this sequence.

LAUNCHING THE TASKS

You might show students a range of different sized rectangular packages and ask what is the same and what is different about them. Alternatively, you might read a picture storybook, such as *The Man Who Loved Boxes* (by Stephen King), as a stimulus for an initial discussion and as an opportunity to elicit language associated with prisms.

ASSESSMENT

Suggested pre- and post-test assessment task: Suggestion 3, '12 cubes'.

The following are some specific statements to inform assessment.

Students are learning about volume when they can:

- create a mental image of the dimensions of a given rectangular prism and build and draw the prism
- interpret the dimensions of length, width and height to construct a prism
- explain how dimensions can be used to find the volume of a prism
- construct different prisms for a given number of cubes and describe similarities and differences among them
- describe a given prism in sufficient detail for someone else to construct sight unseen
- visualise then construct a prism from a description
- recognise and explain that different shaped prisms can have the same volume
- estimate and correctly quantify the number of cubes required to make a prism, without counting by ones

- compare and order prisms using direct comparison based on their volume
- estimate and determine the number of cubes needed to build a non-rectangular prism
- explain and record two different ways of finding solutions
- describe the similarities and differences between rectangular and non-rectangular prisms.

SUGGESTION 1: INVESTIGATING THE VOLUME OF PRISMS

MATHEMATICAL FOCUS

The volume of a prism can be found by determining the number of cubes needed to build it.

PEDAGOGICAL FOCUS

Encourage students to estimate the number of cubes in each prism before working out the exact number, then work out the quantity of cubes required without counting by ones. Building on the language of prisms explored in the 'Objects' sequence, ask students questions such as: *What is the same about these objects? What is different about them?* Such questions will reveal what students are attending to when considering the structure of prisms (e.g. the size, or possibly the language, of the dimensions).

Because the tasks use materials, there is less need for enabling and extending prompts.

For this suggestion, have wooden cubes (2 cm^3) available for students to use, and use small boxes for the consolidating tasks.

HOW MANY CUBES?

How many cubes are needed to make each of these prisms?

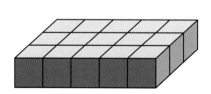 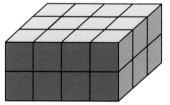 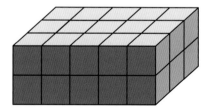

Comment

There is a single solution for the number of cubes required for each of these prisms; what will vary are the strategies students use to determine the quantity. Students may use a solution for one prism to find the solution to one of the other prisms.

CONSOLIDATING THE LEARNING

There are further tasks focusing on determining the total number of cubes required for each prism. They introduce the idea that volume is associated with the space an object takes up and can be measured using informal units such as cubes.

Drawing on the experiences of the previous task, encourage students to estimate the number of cubes they think will be needed. Provide four different shaped boxes, two with the same volume and the other two with different volumes. Make sure that the volumes of the boxes can be found using cubes (e.g. 2 cm^3).

WHAT MIGHT THE PRISM LOOK LIKE?

I made a prism that took up the same amount of space as one of these boxes. How many cubes did I need and what did the prism look like?

 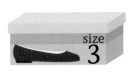  

MYSTERY BOX

I found a box that took up the same amount of space as 20 of these wooden blocks (2 cm cubes). What might my box look like?

Describe each of these prisms to a friend (pretending you are on the phone).

Each of these prisms is different from the others.

Describe each of the prisms so that someone else would understand which one you are referring to.

SUGGESTION 2: CREATING PRISMS FROM A DESCRIPTION

MATHEMATICAL FOCUS

Rectangular prisms can be described by length, width and height.

PEDAGOGICAL FOCUS

Students use wooden cubes to create and interpret descriptions of regular objects (namely prisms) based on how long, how wide and how high they are. Encourage students to imagine what the prisms might look like before they construct them. A key aspect of this suggestion is the language used to describe prisms and to recognise what is the same and different, such as discussing the dimensions of length, width and height.

For this suggestion, have wooden cubes (2 cm^3) available for students to use rather than multilink blocks, as they have nodules. Vocabulary cards can be used when students are describing the prism.

Prior to launching the task, invite students to describe one or more of the prisms from Suggestion 1 to focus on the language. Asking questions such as: *How are they the same?*, *How are they different?* might prompt language such as 'layers', 'rows', 'levels', 'three-dimensional arrays'.

BUILDING A PRISM FROM A DESCRIPTION

Use cubes to build a rectangular prism with a base length of 2 cubes, a width of 2 cubes and 3 layers of cubes high.

Draw what your prism looks like.

Comment

While there is only one solution to this task, part of the discussion could focus on the dimensions, and how many cubes were needed. Although everyone can construct the same prism (i.e. with the same dimensions), what the prism looks like can depend on its orientation.

Enabling prompt

For students needing a prompt associated with the language, ask them to look at the prisms from the previous task and identify their length, width and height.

Extending prompt

To extend the task, ask students what the dimensions might be if they made a different shaped prism using the same amount of cubes.

CONSOLIDATING THE LEARNING

The following tasks can be used to consolidate the learning from the 'Building a prism from a description' task. The emphasis is on interpreting the dimensions of the prism and the number of cubes needed to construct the prism. Again, encourage students to visualise what the prism might look like before construction and to possibly draw it. In both instances, note how well the students are able to interpret the dimensions and deliver the instructions.

BUILDER AND ARCHITECT

One person is the builder and the other is the architect. The architect chooses one of the prisms from the earlier task (describing the prisms to your friend), and records the length, width and height of it. Give these dimensions to the builder to use to construct the prism.

BACK-TO-BACK CONSTRUCTION

In pairs, sitting back to back, each person builds a prism and records the length, width and height. Take turns to describe the dimensions to your partner, who will then build the prism. Check their prism against the original design.

SUGGESTION 3: CREATING DIFFERENT SHAPED PRISMS

MATHEMATICAL FOCUS

Prisms with the same volume can have different dimensions.

PEDAGOGICAL FOCUS

A key aspect of this suggestion is for students to recognise the dimensions of different prisms and that all the prisms (with the same number of cubes) have the same volume. Students might make connections to arrays and multiplicative relationships. Some students will use cubes to create the prisms, whereas others might draw or describe them. Encourage students to imagine their solution before they start the task; you might even suggest they draw the solution they imagined. Students should be encouraged to find more than one solution. Invite students to record and label their solutions and to check for duplicates. Students' recording of solutions can be used for assessment purposes.

12 CUBES

A rectangular prism is made from 12 cubes.
What might the prism look like?
Give some different answers to this.

Comment

There are four different solutions for this task, without duplication (12 by 1 by 1; 6 by 2 by 1; 4 by 3 by 1; 3 by 2 by 2).

Enabling prompts

Build a rectangular prism from 6 cubes.

Extending prompt

Convince me you have found all the possibilities.

CONSOLIDATING THE LEARNING

There is one further task presented that can be used to consolidate the learning. If necessary, you can create similar questions.

MYSTERY PRISM

I used 24 cubes to construct a rectangular prism. What might the prism look like? Give as many different possibilities as you can.

SUGGESTION 4: IMAGINING WHAT THE PRISM MIGHT LOOK LIKE

MATHEMATICAL FOCUS

The dimensions of a prism can be used to create a mental image of what it might look like.

PEDAGOGICAL FOCUS

In the previous suggestions students constructed the prisms. This suggestion instead requires students to imagine or draw what the prism might look like to support them in solving the tasks. Ask students to explain how they used what they had imagined to determine the number of cubes needed to build the prism. Doing so could support other students' learning, such as the language of length, width and height; and noticing layers, which links to arrays and multiplicative thinking, where the number of cubes along each dimension is treated as a unit.

2 LAYERS HIGH

How many cubes are needed to make a rectangular prism that is 2 layers high and has a base 3 cubes wide and 5 cubes long?

Comment

While there is only one solution to this task and subsequent tasks in this suggestion, the focus is on the different strategies that students use to determine the number of cubes needed to make the prism.

Enabling prompt

To assist students with the language of length, width and height, show them a prism and describe it in terms of its dimensions. Next, hide the prism and ask students to describe it, draw it and determine the number of cubes needed to make it.

Alternatively, let students use paper to draw the prism, or use cubes initially with a simpler task, then encourage them to visualise what the prism might look like.

Extending prompt

If the prism is twice as wide, twice as long and twice as high, how many cubes would be needed to build it?

CONSOLIDATING THE LEARNING

The following tasks continue to explore the idea of visualising prisms through imagining their dimensions. To extend learning, students might be encouraged to notice the multiplicative relationship of the dimensions. For example, they might make connections to area (knowing the length and width) to work out the number of cubes in each layer, and then consider how many layers there are.

Pose questions such as, *What might be the dimensions of other prisms made with the same number of cubes (volume)?* to further explore some of the ideas in this suggestion.

5 LAYERS HIGH

How many cubes would be needed to make a rectangular prism that is 5 layers high, has a base 2 cubes wide and is 4 cubes long?

6 LAYERS HIGH

How many cubes would be needed to make a rectangular prism that has a base 10 cubes wide, 10 cubes long, and 6 layers high?

100 LAYERS HIGH

How many cubes would be needed to build a tower in the shape of a rectangular prism that is 100 cubes high, 5 cubes wide and 3 cubes deep?

HOW MANY CUBES?

Roll a dice. That number is how many cubes your prism is long.
Roll the dice again. That number is how many cubes your prism is wide.
Roll the dice again. That number is how many cubes your prism is high.
How many cubes would you need to build your prism?

SUGGESTION 5: INVESTIGATING 'STEPPED' PRISMS

MATHEMATICAL FOCUS

There are different ways to determine how many cubes are needed to build 'stepped' prisms.

PEDAGOGICAL FOCUS

As this suggestion focuses on 'stepped' prisms, it is important to give students some time to describe how these questions are different from questions about rectangular prisms and how they are similar. As students are working, notice whether they use the language of length, width and height to determine the number of cubes.

Visualising the prism as a set of steps, the right-hand step (of eight cubes) can be imagined to move to the top of the middle step, creating a rectangular prism six cubes high, two cubes deep and five cubes wide. This could be part of a discussion in the 'Summarise' phase of the lesson. Doing the task two ways (and getting the same answer) is an important insight for students. Emphasise recording of solutions in ways that communicate students' thinking and encourage them to annotate the diagram as they are working.

When launching the suggestion, you might have a physical model of this prism (made with cubes) and a rectangular prism and ask how they are the same and how they are different. Record some of the language used. Encourage students to imagine their solution before they start the task.

HOW MANY CUBES IN THIS STEPPED PRISM?

Work out your answer in two different ways.

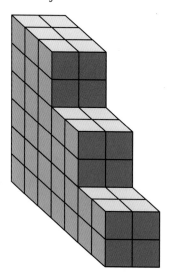

Comment

While there is only one solution to this task, there are different strategies students might use to work out the number of cubes needed to make the prism. Some students might visualise the eight cubes on the right-hand end being put on top of the next step to form a rectangular prism and work out the number of layers (6 layers of 10). Other students might decide to create two or three rectangular prisms by rearranging this larger prism, calculate the number of blocks in each of these smaller prisms, and then add them together. Note whether the students are recording units, such as '60 cubes'.

CONSOLIDATING THE LEARNING

There are further tasks that can be used to consolidate and extend the learning from the 'How many cubes in this stepped prism? (1)' task. The first two tasks ('How many cubes in this stepped prism? (2)' and 'How many cubes in this H prism?') consolidate the learning from the first one. Encourage students to think about using what they know to work out the number of cubes and to explain two different ways they arrived at their solution. The last two tasks in this suggestion ('An L shape' and 'A U shape') extend the learning, as the students need to construct multiple prisms of a certain volume (number of cubes).

HOW MANY CUBES IN THIS STEPPED PRISM? ❷

Work out your answer in two different ways.

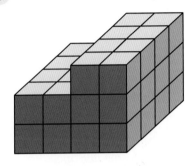

HOW MANY CUBES IN THIS H PRISM?

Work out your answer in two different ways.

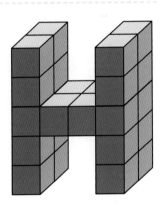

AN L SHAPE

Use 30 cubes to make a shape like an "L". Draw or describe how you did it. Do this in two different ways.

A U SHAPE

Use 36 cubes to make a shape like a (block) "U". Draw or describe how you did it. Do this in two different ways.

Comment

After the students have worked through the different tasks, have a discussion about what is the same and what is different about these prisms and what students think they were measuring.

CHAPTER 10

RECOGNISING POLYGONS

OVERVIEW

This sequence is suitable for students from Foundation to Year 2. Suggestions 1 and 2 are suitable for Foundation to Year 2 students, whereas Suggestions 3 and 4 are suitable for Years 1 and 2. The suggestions offer experiences of classifying, making, naming and describing two-dimensional polygons.

When teaching this sequence, we use the term 'polygon' when referring to closed figures with at least three straight sides. The term 'shapes' is common; however, it can be misleading because not all shapes are polygons. The focus of student learning includes prompting them to imagine, classify and describe attributes of polygons.

This sequence includes ideal preliminary experiences to the 'Reasoning with polygons' sequence.

The following is a summary of the suggestions in this sequence.

	MATHEMATICAL FOCUS
Sorting and classifying polygons	Polygons are closed shapes made from straight lines which can be sorted according to their mathematical properties.
Investigating triangles	Different triangles can have similar properties but can also be different.
Making and naming polygons	Polygons can be made, named and drawn according to their properties.
What might be the polygon?	Geometric language can be used to describe what we imagine but cannot see.

RATIONALE

The focus of this sequence is on exploring properties of polygons. Teachers will help students to develop vocabulary and identify key features of different polygons. These experiences lay the foundations for geometric reasoning as students justify their thinking, explore conjectures and develop logical reasoning skills.

Prior to teaching the tasks, it is important that teachers explore the solutions. This will include considering and learning the names of different polygons, their properties and relationships.

Some suggestions have enabling prompts, but there are few extending prompts because the main tasks are open ended, having more than one solution.

LANGUAGE

triangle, trapezium, square, parallelogram, quadrilateral, polygon, pentagon, octagon, kite, hexagon, equal length sides, straight sides, vertex, vertices and closed figures

LAUNCHING THE TASKS

The following is a possible game suitable for launching various tasks.

POLYGON CELEBRITY HEADS

BLM 16

Students wear polygon cards on their head. They ask their classmates questions about the properties of each polygon and guess which polygon they are wearing.

Adapt the cards to the focus of the task for the lesson. A sample set of cards is provided on **BLM 16: Polygons celebrity heads**.

Oxford OWL

ASSESSMENT

BLM 17

Suggested for pre- and post-assessment task: **BLM 17: Naming polygons**.

Sort the figures into two groups: those you are sure are triangles and those that might not be triangles. Explain how you made your decisions.

The following are some specific statements to inform assessment.
Students are learning about polygons when they can:
- sort polygons into groups, and describe properties they have used for sorting
- classify polygons in various ways
- recognise polygons as closed figures made from straight lines
- construct, draw and name polygons that they create
- describe properties of polygons they make.

Oxford OWL

SUGGESTION 1: SORTING AND CLASSIFYING POLYGONS

MATHEMATICAL FOCUS

Polygons are closed shapes made from straight lines which can be sorted according to their mathematical properties.

PEDAGOGICAL FOCUS

Students are encouraged to notice and identify polygons and non-polygons. Polygons are any two-dimensional closed shape with straight lines. Students begin to reason about different groupings of polygons which can be sorted into categories and named by the number of sides and corners. Students will name sets according to the properties of each group. It is important that students explain and justify how they have sorted their groups, and begin to notice and describe mathematical properties of the polygons.

BLM 18

Oxford
OWL

Before beginning the 'Sorting shapes' activity, provide students with a copy of **BLM 18: Sorting shapes**.

There are no enabling or extending prompts, as there are many different ways the images can be sorted.

SORTING SHAPES

Can you sort these shapes into groups?

Describe the groups.

What might be another way to sort these shapes?

Can you sort these shapes another way?

Comment

Examples of categories might include closed or not closed, number of sides, corners or no corners.

CONSOLIDATING THE LEARNING

There are five further tasks that could be used to consolidate the learning. The first three use pattern blocks. Students may sort by colour, number of sides, size of polygon, or quadrilaterals or non-quadrilaterals.

There are 3 secret sort tasks. The solutions for each sort are: 'Secret sort 1': 4-sided figures/ trapeziums; 'Secret sort 2': Polygons/triangles; 'Secret sort 3': pentagon/hexagons.

SORTING POLYGONS ❶

Sort these polygons into 2 groups.

Record your answer and describe each group.

Can you sort them a different way?

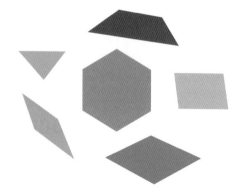

SORTING POLYGONS ❷

Sort these polygons into 3 groups, 4 groups and 5 groups.

Record your answers and describe each group.

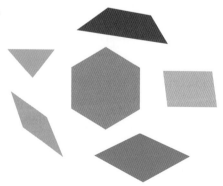

These sets of polygons have been sorted into groups. Explain why they may have been grouped together in this way.

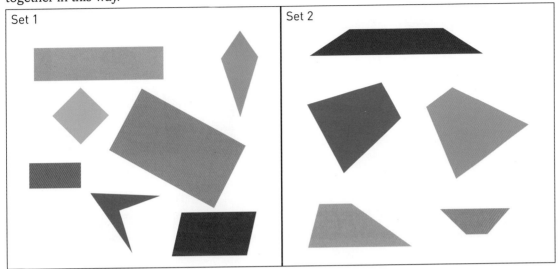

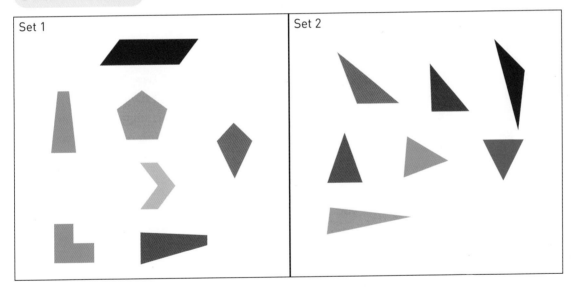

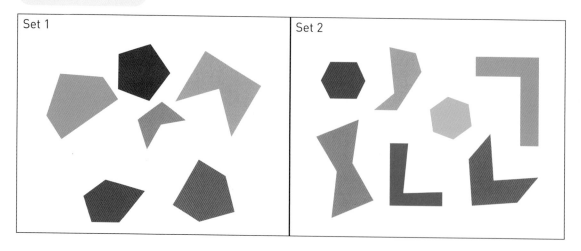

| Set 1 | Set 2 |

SUGGESTION 2: INVESTIGATING TRIANGLES

MATHEMATICAL FOCUS

Different triangles can have similar properties but can also be different.

PEDAGOGICAL FOCUS

BLM 19

Oxford OWL

Not all shapes made from three lines are triangles. A triangle is a closed, two-dimensional polygon with three straight sides. One of the consolidating tasks has a range of polygons, some of which are triangles and some of which are not. The emphasis in all tasks in this suggestion is on student explanations of their reasoning. Give students a copy of **BLM 19: Which ones are triangles?**

TRIANGLES ON DOT PAPER

Draw as many different triangles as you can by connecting three dots on isometric paper. Explain how your triangles are different or the same.

Comment

Triangles can differ by their properties: some have all sides equal (equilateral), some have two sides equal and the other side different (isosceles), some triangles have a right angle (right angled), and some have all sides different (scalene). Triangles can also have the same shape but be different in size. Encourage students to find as many different triangles as they can, to describe the triangles they draw and to explain their thinking.

Enabling prompt

What is the smallest triangle you can draw on the isometric paper with three lines drawn from dot to dot?

Extending prompt

Draw two triangles that are the same but the lines of one triangle are twice as big as the lines of the other.

CONSOLIDATING THE LEARNING

There are three further tasks that can be used to consolidate the learning from the first one. The second task requires 2 metre lengths of rope (or string) per group of students. Once the triangles are made, have each group of students place theirs on the floor for all to see (still holding the rope in place). Capture some photos of the different triangles to use as part of the discussion.

TRIANGLES WITH STICKS

Make some triangles using materials such as GeoStix. In what ways are your triangles similar? In what ways are they different?

TRIANGLES WITH STRING

Make some triangles using string. In what ways are your triangles similar? In what ways are they different?

WHICH ONES ARE TRIANGLES?

Provide students with a copy of **BLM 19: Which ones are triangles?**

Sort the shapes on the worksheet into two groups: those you are sure are triangles and those that might or might not be triangles.

Explain how you decided on your two groups. What is the same about each of the groups? What is different?

BLM 19

Oxford
OWL

SUGGESTION 3: MAKING AND NAMING POLYGONS

MATHEMATICAL FOCUS

Polygons can be made, named and drawn according to their properties.

PEDAGOGICAL FOCUS

Students make polygons then compare the sides to classify their polygons.

Prior to launching the task, it is recommended that students play with the geoboards if they have not used these before. They might make their own patterns, pictures or polygons and share these as a class.

Some geoboards are two sided: squares on one side and triangles on the other (see Figure 2). The boards with spikes in the shapes of triangles are named 'isometric geoboards' and are used for this suggestion. As an alternative, you might download a geoboard app.

Figure 2   Double-sided geoboards (square and isometric)

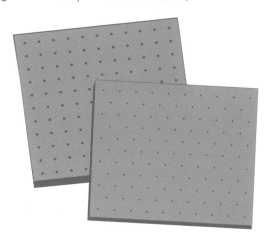

The intention is that students make polygons on isometric geoboards and/or draw them on isometric paper. Some students may easily use the dot paper, whereas other students may need to make the polygons on geoboards. Encourage students to visualise prior to making or drawing the polygons.

Students might work in pairs. Encourage students to make different polygons (including irregular ones) and notice what is the same or different about their polygons (focusing on corners [vertices] and sides).

When a polygon is rotated or flipped, it is the same polygon.

A focus in many of the tasks is to encourage students to describe the polygons they make or draw.

MAKING AND DRAWING POLYGONS

Make as many different polygons as you can.

Draw each of your polygons on isometric dot paper and write the name of your polygons and anything else you notice about it.

Enabling prompt

Make a triangle on the geoboard and draw it on the isometric paper.

Extending prompt

A concave polygon goes inwards like a cave. Make some different concave polygons and draw them on the isometric paper. Explain in what ways they are different.

CONSOLIDATING THE LEARNING

There are further tasks that can be used to consolidate the learning from the first task.

OXFORD UNIVERSITY PRESS

MAKING POLYGONS OUT OF TRIANGLES

If you have 6 triangles (all the same), what polygons can you make using some or all of the triangles?

Visualise what the polygon might look like.

Draw your polygons on isometric dot paper and name them.

DESCRIBE AND DRAW A POLYGON

Imagine you are talking to your friend on the phone.

Describe one of the polygons you made using six triangles.

Your friend draws what you describe.

Check that your friend's polygon is the same as yours. Explain ways that their polygon is the same as yours.

Now let your friend try.

MAKING POLYGONS OUT OF RHOMBUSES

If you have four rhombuses (all the same), what polygons can you make?

Draw your polygons on isometric dot paper.

Name the polygons.

DESCRIBING POLYGONS MADE USING RHOMBUSES

Assume you are talking to someone on a mobile phone.

Describe one of the polygons you just made.

The listener(s) draw what you describe.

Explain ways that their polygon is the same as, or different from, yours.

POLYGONS WITH STRING

Each group of students has string tied in a loop. With each person holding the string tight, make and draw a:

- trapezium
- parallelogram
- rhombus
- square
- quadrilateral that is different from any of the other polygons.

Comment

After students have made each of the polygons, capture some photos to use as part of the discussion. Ask them: *What is the same about each of the polygons? What is different about the polygons?*

There are four 'Which one might not belong?' tasks created but students may wish to create their own and test a partner. There is more than one possible solution for each of these tasks.

Encourage students to explain their reasons for their choice and to record their thinking. The class discussion during the 'Summarise' phase provides an opportunity to focus on the properties of the different polygons and the relationships between them.

Set 1: Which one might not belong? Explain your reasons.

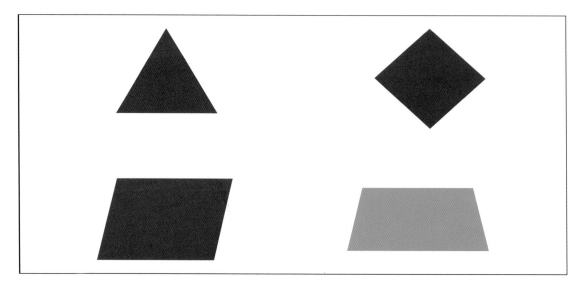

Set 2: Which one might not belong? Explain your reasons.

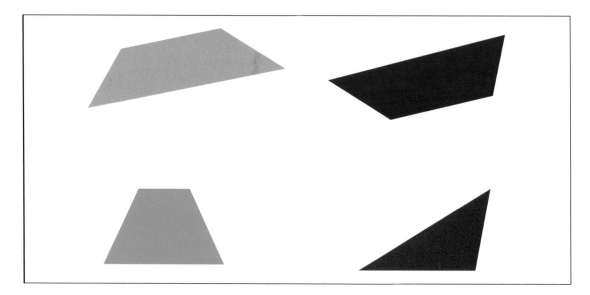

Set 3: Which one might not belong? Explain your reasons.

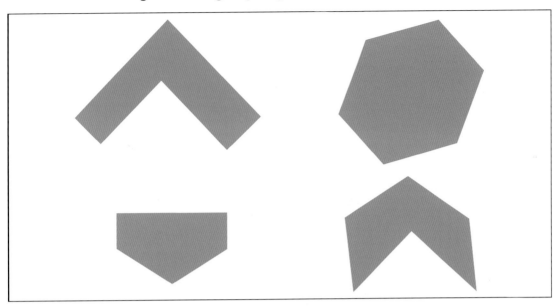

Set 4: Which one might not belong? Explain your reasons.

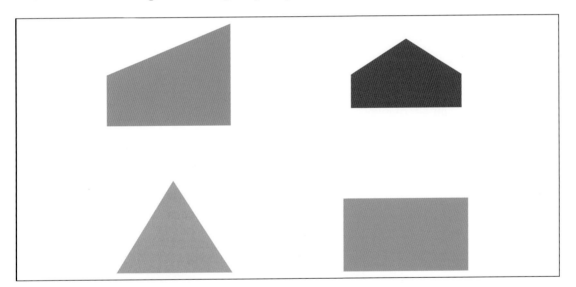

SUGGESTION 4: WHAT MIGHT BE THE POLYGON?

MATHEMATICAL FOCUS

Geometric language can be used to describe what we imagine but cannot see.

PEDAGOGICAL FOCUS

The focus is on the students being able to speculate, describe and justify what they think is hidden. Ask probing questions such as, *What did you picture in your mind? What polygons did you know it could not be and why?* After students find one possibility, encourage them to find others.

Once they have found a range of possibilities, have students display them on the floor (especially if using geoboards for recording). Ask, *What do you notice about all the possibilities? Are there any other possibilities? Why not?*

IMAGINING POLYGONS ❶

Someone drew a polygon but a page fell on their work and covered part of the polygon. What polygon might they have drawn?

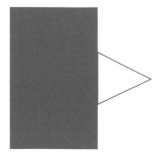

Comment

There are a number of possibilities for this task, including a triangle, parallelogram, trapezium, or rhombus. A key aspect of this task includes students justifying their solutions and comparing dimensions of the polygons.

Enabling prompt

Show the students an alternative image such as a rectangle hidden under the paper and ask, *What might the hidden polygon be?*
Trace with your finger to show what the hidden part of the polygon might look like.

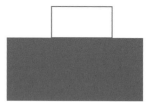

As an additional prompt, have some pattern blocks on the table and ask the student which of these polygons could be the hidden polygon. *Why?*

Extending prompt

Convince me that you have found all possibilities. How did you decide which polygon it could be?

CONSOLIDATING THE LEARNING

There are three further tasks that can be used to consolidate the learning from the first task. The students could also work in pairs, with one child hiding a polygon and the other imagining what the polygon might be.

IMAGINING POLYGONS ❷

What other polygons might the person have drawn?
What do you notice about the properties of all these polygons?
Which polygons do you know are not possible and why?

My friend drew a polygon and then accidentally knocked some paint onto her work. Imagine what the polygon might have looked like, and then draw it. Is there more than one possibility? Why? Why not?

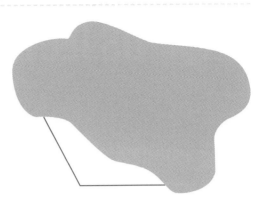

CHAPTER 11

OBJECTS

OVERVIEW

This sequence is suitable for students from Foundation to Year 2. Suggestions 3 and 4 are more suitable for Years 1 to 2. Some of the content is beyond what is stipulated in the curriculum, but is intended to connect with ideas in the other geometry sequences and the volume sequence. It is recommended that students engage in the 'Recognising polygons' sequence prior to this sequence.

The following is a summary of the suggestions in this sequence.

	MATHEMATICAL FOCUS
Three-dimensional objects in everyday life	Spatial language can be used to describe everyday three-dimensional objects.
Properties of three-dimensions objects	Three-dimensional objects can be sorted and classified in many ways.
2D shapes within 3D objects	Relationships between two-dimensional shapes and three-dimensional objects can be explored using materials.

RATIONALE

We live in a three-dimensional world, and children come to school having had many experiences exploring a range of objects through play and construction. It is essential that we build on these experiences and provide opportunities for students to develop an understanding of three-dimensional objects that goes beyond simply naming and describing objects, to recognising the relationship between two-dimensional polygons and three-dimensional objects, as well as the functionality of objects in everyday life. Opportunities for students to explain and justify these relationships (such as cones and pyramids) help to deepen their understanding.

The goal of this sequence is to offer students experiences of manipulating, classifying, making, naming and describing three-dimensional objects. In doing so, students are developing an understanding of the properties of objects, of geometric language used to describe them, and of the relationship between polygons and objects. The ideas in this sequence and the other geometry sequences lay the foundation for the geometric knowledge and spatial reasoning (which includes visualisation, mental rotation and spatial language) that students will need in other aspects of mathematics, such as number and measurement.

RESOURCES FOR THIS SEQUENCE

Before commencing this sequence, we recommend collecting everyday items of different three-dimensional objects (e.g. prisms, cylinders, cones, spheres, pyramids) and taking photos of them to replace the exemplar images in this sequence. Encouraging students to bring objects from their home and setting up a table in the classroom can enrich the experiences and serve as resources when exploring the tasks.

Other resources for this sequence include: sticks (or straws or craft sticks) and Blu Tack, small cubes, polydron materials, multilink blocks (for Suggestion 5, if the cubes are too difficult for students to manipulate), and three-dimensional solids. A collection of different and interesting shaped boxes and everyday objects is also recommended.

At all levels, teachers have found that stories have provided a useful context to connect with the functionality of three-dimensional objects as well as their structure. Examples of books that might be useful are *The Man Who Loved Boxes* (Stephen Michael King); *My Cat Likes to Hide in Boxes* (Eve Sutton); *I Can Build a House* (Shigeo Watanabe).

LANGUAGE

Three-dimensional: cone, cube, cylinder, pyramid, prism, rectangular prism, sphere, square based pyramid, torus (doughnut-shape), edges, vertices (corners), faces, curved, flat, base, polyhedron

Everyday language: ball, box, cone, cube, dice, tube, can, wedge, egg, doughnut, solid, hollow, points, corners

Functional language: roll, slide, stack, holds liquid, storage container

Two-dimensional: triangle, square, rectangle, parallelogram, quadrilaterals, pentagons, hexagon, octagon, side, congruent, polygon

LAUNCHING THE TASKS

Prior to teaching this sequence, you could prepare sets of laminated cards with words from the 'Language' section. Use the cards to launch the tasks in this sequence.

Using a collection of everyday objects (e.g. tissue box, basketball, drink bottle, dice) for students to touch and describe, ask questions such as, *What do you notice about these objects?* and *What do you wonder about…?*. Such questions can be used to gain insights into the language students use and what they attend to, such as functionality.

ASSESSMENT

The objects sequence comprises various objects; thus, it is important to consider the concept you intend to assess. For example, naming and describing objects (suggested pre-test and post-test assessment task: Suggestion 1: 'Same and different (1) and (2)').

The following are some specific statements to inform assessment.
Students are learning about three-dimensional objects when they can:
- name and describe three-dimensional objects in everyday life using everyday language and identify which are the same shape but have different everyday uses

- identify, describe and explain the similarities and differences of objects using everyday language and spatial language
- recognise that we can represent a three-dimensional object as a two-dimensional picture or diagram
- sort objects into two or more groups and justify the categories using spatial language
- correctly name an unseen object from an oral or written description of its properties
- identify the two-dimensional faces of three-dimensional objects and name the polygons of these faces
- construct, name and describe objects using spatial language
- pose questions or statements to describe the relationship between two-dimensional shapes and three-dimensional objects
- visualise, predict, assemble and describe different constructions of three-dimensional objects made with small cubes.

SUGGESTION 1: THREE-DIMENSIONAL OBJECTS IN EVERYDAY LIFE

MATHEMATICAL FOCUS

Spatial language can be used to describe everyday three-dimensional objects.

PEDAGOGICAL FOCUS

The focus in this suggestion is on developing the language associated with three-dimensional objects and identifying similarities and differences between them. Whilst Tasks 1 and 3 focus on describing everyday objects, Tasks 2 and 4 relate to functionality as well as shape. Encourage students to record their thinking in some way. For example, students could draw a table, record their thinking on a worksheet, or label pictures. Pose questions to challenge student thinking and reasoning, such as, *I think the pyramid and cone are alike. Why would I think this?* or *I think the drink bottle and brick are different. Why would I think this?* Students could make some sort of ramp for the fourth task to explore the function of the objects.

When launching this task, have a class discussion about what students know about the different everyday objects and build up a list of words they use to describe them. Record these words on a poster for students to refer to later. Students could take photographs of three-dimensional objects in their lives, and share these with the class to build an awareness of the different objects and their functionality in everyday environments.

Further, have a collection of everyday objects on the floor for students to explore. The intention is for them to discuss everything they know about the objects and then to share with the whole group. You might ask students if they could find something in the room that is like their object in some way. Record their descriptions on a poster to refer to throughout the sequence.

What are some names we give these everyday objects?

CONSOLIDATING THE LEARNING

There are three further tasks that can be used to consolidate the learning from this first task. However, 'Roll, stack and slide' is different, as the focus is on the functional aspect of the objects.

SAME AND DIFFERENT ①

Which of these objects are alike in some way?

Which of these objects are alike in some way?

ROLL, STACK AND SLIDE

Find three different objects in the room that you think can roll.

Now find three different objects in the room that you think can stack.

Are there some objects in the room that you think can slide?

Share your findings with a partner. How might you test them out?

SUGGESTION 2: PROPERTIES OF THREE-DIMENSIONAL OBJECTS

MATHEMATICAL FOCUS

Three-dimensional objects can be sorted and classified in many ways.

PEDAGOGICAL FOCUS

This suggestion focuses on sorting and classifying as a vehicle for developing the spatial language associated with three-dimensional objects and identifying similarities and differences among them. As was the case in the previous suggestion, it is important to have the physical objects for students to use, and to focus on the language they are using to justify how they sorted the objects.

For Task 1, 'Secret sorting', give students a few minutes to explore the objects and then to sort them in some way. It is advisable to have someone record the criteria for sorting. Then they invite another pair (or group) to decide how they sorted them. Extend by asking them to sort the objects into more than two groups. The third task, 'What am I?', focuses on students identifying an unseen object that is being described to them, and so requires them to interpret the language and visualise what the object might be. If doing this task as a whole class with younger students, it is best to have a collection of five objects on the floor for them to see, and an identical set in the bag. Present the poster with key language previously recorded to refer to, and record the questions asked, along with student responses. After about three questions, ask: *Which objects do we know it could not be? Why?*

When launching this suggestion, you might have four or five objects on the floor and pose the questions, *I sorted these objects into two groups. How might I have sorted them? What might my secret rule be?* Invite students to share their suggestions and record the responses. Give others time to consider whether the suggestion holds for both groups of objects, or if they want to challenge or modify it. Have the secret rule recorded on cards (e.g. more than one rectangular face, no rectangular faces) to display once students have agreed on the criteria. Ask questions such as, *What other objects might belong in each group? Why?* Students then work in pairs or small groups selecting ways to sort their collection of objects.

SECRET SORTING

Sort these objects into groups.
Ask a friend to guess how you sorted them.
How else might you sort the objects?
Now ask your friend to sort them.

Comment

There are several solutions to the first task, but a key aspect is for students to explain the criteria of how others have sorted the objects, using spatial language. Record some of the criteria they suggested on cards, for later use.

Enabling prompt

For students who might require a prompt, invite them to choose two objects that are alike in some way and to explain how they are alike (e.g. pointy top). Record their response on a card. Then ask them to find two that are alike in a different way and again describe and record this (e.g. no pointy top).

Extending prompt

Present a Venn diagram, such as the one shown below, to the students. Ask them to suggest what the criteria might be for the red and blue regions and what objects might go in the middle, and why.

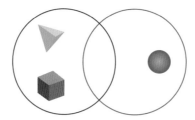

Record the criteria for each of the regions. For example, objects in the red region could have more than one flat surface, whereas objects in the blue region can roll. Objects in the overlapped region could have one flat surface and can roll, such as a cylinder or cone.

Students could then create their own Venn diagram for others to solve.

CONSOLIDATING THE LEARNING

There are two further tasks that can be used to consolidate the learning following the first task. The second task, 'What am I?', focuses on spatial visualisation as well as spatial reasoning. You might do this task as a whole class first and record the questions. After two or three questions ask the students to draw what they think the object might be, and provide a reason for this.

WHICH ONE MIGHT NOT BELONG?

Which one of these objects do you think might not belong in this group?
Why not?

OXFORD UNIVERSITY PRESS

WHAT AM I?

This bag has some objects in it.

In small groups, one student feels and describes an object. The other students take turns to ask questions that the first student can only answer with 'yes', 'no', or 'not sure'.

Comment

The aim is to name and draw the object, and to justify their reason, before the object is revealed.

SUGGESTION 3: 2D SHAPES WITHIN 3D OBJECTS

MATHEMATICAL FOCUS

Relationships between two-dimensional shapes and three-dimensional objects can be explored using materials.

PEDAGOGICAL FOCUS

A key focus of the tasks in this suggestion is for students to recognise that in many three-dimensional objects, the faces consist of two-dimensional shapes. This understanding needs to develop over the course of this suggestion and subsequent suggestions. Students will require independent thinking time to consider possibilities, and to record them with a justification.

For this suggestion, have a range of interesting boxes (not just rectangular and square prisms) for pairs of students to explore, key language cards, A3 paper, three-dimensional solids (prisms and pyramids), sticks (craft sticks or straws) and Blu Tack. This suggestion is intended to go across two or more lessons.

Throughout the suggestion, encourage students to discuss similarities and differences between objects, and record their ideas on a class chart. Doing so focuses students' attention on objects that are polyhedrons (three-dimensional objects made of polygons such as prisms and pyramids), and other three-dimensional objects (cones, cylinders and spheres). Such discussions also reveal what students are noticing and the relationships they are recognising. Similarly, encouraging students to listen to what others are saying provides them with the opportunity to expand their own perceptions of three-dimensional objects.

Prior to launching the tasks, have a collection of boxes in the middle of a circle of students. Begin with a game of 'I spy the faces of the objects' to tune them into the language. *I spy a box with a curved face. Which box might it be? What is a mathematical name for this box?* While playing the game, record the language on the board and use the vocabulary cards. Invite the student who selected the correct box to have a turn. After 5–6 turns ask, *What do you notice about the faces of these boxes?* (Making the connection to two-dimensional shapes.)

The following are some pedagogical considerations when implementing the consolidating tasks. It is important that students can explain their understanding of edges and corners using their own language, and can find examples of these in the classroom. Questions to focus their attention include: *What do you notice about the number of edges and corners of each object? Which skeletons are alike in some way? Which ones are different? What is the same about these objects? What is different?*

MYSTERY BOX

I traced the faces of these boxes onto paper, then I put the box back in the collection.

I traced six faces; some of them were rectangles. What might be the mystery box?

Comment

There are several solutions to the first task, but a key aspect is for students to recognise and describe the faces of the different boxes and justify their choice. Encourage them to record their justification on the sheet of A3 paper. Their recordings can be used for assessment.

Enabling prompt

For students who need a prompt associated with some of the language, have a collection of boxes in front of them. Ask them to choose a box and show you one of the faces. What shape are the faces of the box?

OXFORD UNIVERSITY PRESS

Extending prompt

Include some cones and pyramids and ask students:

- Can you find some different objects that have one or more rectangular faces and faces that are another shape?
- How many objects can you find and what are their names?

CONSOLIDATING THE LEARNING

There are further tasks that can be used to consolidate the learning following the first task. For the 'Lost block' task, asking students to draw the lost block another way can help them see that triangular prisms are still called triangular prisms even when they have different dimensions, or are in different orientations. The clue students use in the 'Secret object' could be used for assessing their language about shape. 'Constructing skeletons' and 'Mystery object' involve constructing prisms and pyramids (polyhedrons) using materials and focusing on the objects' properties: the names and number of faces, number of edges and corners (vertices), and identifying what is the same and different about them.

LOST BLOCK

Lena lost her favourite block. At least one face was a triangle, and at least one other face was a rectangle. What might her favourite block have been?

Draw what the lost block might look like. What might be another possibility?

What is the same about your drawings?

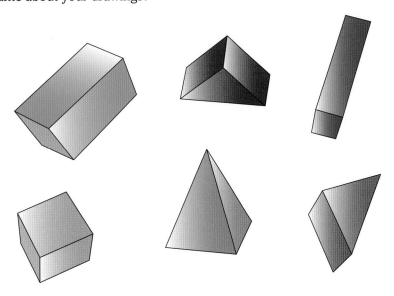

SECRET OBJECT

From a collection of objects, choose one as your secret object.

Write down three clues about your object without saying what the object is. Give the clues to your partner and have them guess which object it is. Ask them to draw the object and to explain which clues helped them work it out.

BLM 20

Oxford
OWL

3D OBJECT FACE MATCH

Give students a copy of **BLM 20: 3D object face match**.

Match the 2D face to the 3D object. On a blank card, draw another 2D face that is on your 3D object. Justify the match.

CONSTRUCTING SKELETONS

Using sticks and Blu Tack, construct as many skeletons of 3D objects as you can.

Describe them to your partner. What is the same about them? What is different?

MYSTERY OBJECT

I used sticks and Blu Tack to construct an object that has six faces. What might it be?

How many different objects could it be?

What might it look like?

Using four small cubes touching each other face to face, make some different 3D objects.

How many different possibilities are there?

OXFORD UNIVERSITY PRESS

CHAPTER 12

REASONING WITH POLYGONS

OVERVIEW

This sequence is suitable for students in Years 1 and 2. It is suggested that students complete the 'Recognising polygons' sequence prior to this sequence.

The goal of this sequence is to offer students experiences at generating new polygons from other polygons (a polygon is a two-dimensional closed shape with straight sides) in two different ways. The first is by combining polygons to make new and different polygons, and the second involves decomposing polygons by making one straight line cut. In doing so, students are building spatial-thinking skills including mental rotation and visualisation, and experiencing polygons in different orientations. Encouraging students to explain and justify their thinking deepens their understanding of the properties of polygons and the relationships between them.

The sequence takes a different perspective from common approaches to learning about polygons in the early years. Rather than relying on repeating definitions, students are invited to solve problems that they have not already been told how to solve.

The following is a summary of the suggestions in this sequence.

	MATHEMATICAL FOCUS
Making polygons using trapeziums	Different polygons can be made by combining trapeziums.
Combining polygons	Polygons created from other polygons can be described and named.
Making polygons from a trapezium	Trapeziums can be split into other polygons.
Making polygons from other quadrilaterals	Other quadrilaterals can be decomposed to make polygons.
Making polygons from other polygons	Polygons can be split and combined to make different polygons.

RATIONALE

Geometry and spatial thinking provide the foundation for learning mathematics and other subjects, and for understanding the spatial world in which we live. Many everyday activities require visualisation, an aspect of spatial reasoning that relates to imagining polygons and three-dimensional objects by creating and manipulating mental images. Teachers can provide learning experiences that develop students' spatial reasoning through active meaning-making.

Within this sequence, the focus is on developing students' understanding of the properties of polygons, recognising relationships between polygons, extending geometric reasoning skills through composing and decomposing polygons to create other polygons, describing, and visualising. The ideas in this sequence build on those in the 'Recognising polygons' sequence and lay the foundation for generalising the properties of polygons using deductive reasoning and the relationship between different polygons.

Due to the tasks having both low floors and high ceilings, there are few enabling and extending prompts suggested.

LANGUAGE

triangle, trapezium, square, rectangle, rhombus, parallelogram, kite, quadrilateral, pentagon, octagon, hexagon, equal length sides, straight sides, parallel sides, angles, right angles, vertex, vertices, polygon, two-dimensional shapes, closed figures

LAUNCHING THE TASKS

Prior to teaching this sequence, we suggest preparing sets of laminated cards with words from the 'Language' section to use throughout the sequence. Pattern blocks and tangrams are also useful for initiating discussions about polygons that incorporate the language associated with them.

When launching the tasks in this sequence, we suggest creating or drawing a set of shape cards such as those shown in Set 1 below. You could ask students to look at each of the polygons and choose those that have the same name, and discuss why they chose the ones they thought of as being the same.

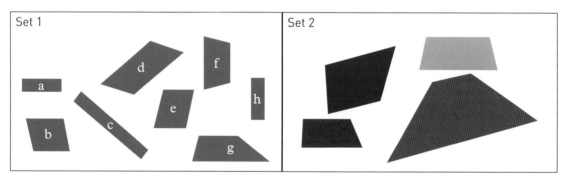

You could also use polygons such as those in Set 2, which are ideal for Suggestion 1. Similar sets can be created for the other suggestions.

Extending on the above idea, you could ask the students to look at the polygons in Set 1 and select one that they think could be the odd one out, explaining why. The same could be asked of Set 2.

Another option is to use one of the 'Secret sort' tasks from Suggestion 1 of the 'Recognising polygons' sequence.

ASSESSMENT

Suggested pre-test and post-test assessment task: Suggestion 4: 'Making polygons from other quadrilaterals'.

The following are some specific statements to inform assessment.
Students are learning about and reasoning with polygons when they can:
• describe the properties of trapeziums, parallelograms and rectangles
• identify and describe similarities and differences between trapeziums, parallelograms and rectangles

- identify and describe similarities and differences between various polygons, including quadrilaterals
- draw polygons to match their definitions
- solve problems involving combining polygons
- solve problems involving splitting polygons from given a polygon
- record possible solutions systematically
- justify reasoning about polygons, especially about properties.

SUGGESTION 1: MAKING POLYGONS USING TRAPEZIUMS

MATHEMATICAL FOCUS

Different polygons can be made by combining trapeziums.

PEDAGOGICAL FOCUS

The purpose of this suggestion is for students to use spatial and geometric reasoning skills to make, name and describe polygons created by combining trapeziums, and to become familiar with how trapeziums can join together without gaps (that is, tessellate) to make a chevron.

Encourage students to imagine their solution(s) before creating with the materials. They then draw and label their solutions, and explain how they combined trapeziums to create the new polygons. Notice the language used when students explain how they combined trapeziums, such as flip (reflect), slide (translate) and turn (rotate). Focusing on the similarities and differences between the different polygons created will assist students to recognise the relationships between them.

Prior to launching the task, we suggest introducing the chevron. Use prompts such as:

- What can you tell me about the chevron?
- Draw a chevron on isometric paper.
- Now draw your chevron facing in the opposite direction.
- Now draw a chevron that is bigger than your first one.

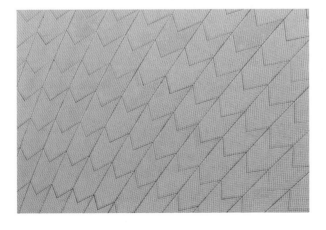

Introduce students to the design of the sails on the Sydney Opera House, which use chevrons. Since opening in 1973, the Sydney Opera House has been one of Australia's most famous and distinctive buildings and was listed as a World Heritage Site in 2007.

You might show students a photo and ask them if they have visited the Sydney Opera House. Ask them to describe what it looks like. From a distance, the sails look white, but close up you can see a pattern; the pattern consists of tessellating chevrons.

MAKING CHEVRONS

Using any of the pattern blocks, make as many different chevrons as you can. Record your chevrons on isometric paper.

What is the same and what is different about all the chevrons you have made?

Comment

It is possible to make chevrons out of parallelograms, trapeziums, triangles and combinations of those. After students have made some chevrons, you might like to ask them: *Why do you think it is possible to make chevrons out of parallelograms, trapeziums and triangles? Is it possible to create a chevron using rhombuses? Why? Why not?*

Enabling prompt

Use four trapeziums to make a chevron.

Extending prompt

Can you make a chevron with an odd number of trapeziums? Convince me.

CONSOLIDATING THE LEARNING

There are two further tasks that can be used to consolidate the learning from the first task, and extend students' experiences of making polygons with trapeziums.

MAKING POLYGONS OUT OF TRAPEZIUMS

Using some or all of four trapeziums (all the same), what polygons can you make?

Make your polygons on an isometric geoboard and/or draw the new polygons on isometric dot paper.

OXFORD UNIVERSITY PRESS

Name your new polygons.

How are your new polygons the same? How are they different?

DESCRIBING POLYGONS MADE OUT OF TRAPEZIUMS

Imagine you are talking to your friend on the phone.

Describe one of the polygons you just made.

Your friend draws what you describe.

Check that your friend's polygon is the same as yours.

Now let your friend try.

SUGGESTION 2: COMBINING POLYGONS

MATHEMATICAL FOCUS

Polygons created from other polygons can be described and named.

PEDAGOGICAL FOCUS

The purpose of this suggestion is for students to use their spatial and geometric reasoning skills to make, name and describe polygons. Provide students with sets of polygons (pattern blocks are ideal) to make regular and irregular polygons. There are many different solutions.

Encourage students to imagine their solution(s) before constructing with the materials, then draw and label their solutions. Invite students to explain how they combined the different polygons to create the new polygons. Having them focus on how their polygons might be the same in some way, or might be different, and to justify their reasons, will assist them to recognise the properties of the different polygons and the relationship between them.

Notice the language the students are using and help them to extend their vocabulary when sharing their answers and thinking.

MAKING POLYGONS

You have four triangles, two rhombuses and two trapeziums.

Draw and name some polygons that you can make using some or all of those polygons.

Comment

It is preferable for students to use isometric paper to draw their shapes, although not essential.

Enabling prompt

Use four triangles to make a polygon.

Extending prompt

How many different polygons can you make? How do you know they are different? Convince me.

CONSOLIDATING THE LEARNING

The following tasks are designed to consolidate students' understanding of making, drawing and naming polygons.

MAKING MORE POLYGONS

You have some triangles, some rhombuses and some trapeziums.

Draw and name some polygons that you can make.

Name and describe your polygons.

How are they the same? How are they different?

MYSTERY BAG

In my bag I have six pattern blocks.

Some blocks have the same number of sides.

Using these blocks, I can make a parallelogram, trapezium, hexagon and a large triangle.

What pattern blocks might be in my bag? Why do you think this?

WHAT MIGHT BE THE NEW POLYGON?

A trapezium, a rhombus and a triangle are combined to create a different polygon.

What might the polygon look like, and how many sides and corners does it have?

How many different possibilities can you find?

What do you notice about them?

SUGGESTION 3: MAKING POLYGONS FROM A TRAPEZIUM

MATHEMATICAL FOCUS

Trapeziums can be split into other polygons.

PEDAGOGICAL FOCUS

A key component of the tasks in this suggestion is visualisation, students' explanations of decisions for cutting the trapezium, and building their knowledge of the properties of resulting polygons. This suggestion also provides students with opportunities to encounter triangles and quadrilaterals with different properties. Taking time to discuss similarities and differences of polygons can expand students' knowledge of polygons beyond those which are regular and vertically facing.

The key focus is on student decision-making: where to draw the straight line on a polygon to create two smaller polygons, and finding multiple possible solutions. Encourage students to imagine their solution before they start the task, justify why they decided to draw the line where they did, and describe similarities and differences between the polygons they made using the correct language.

OXFORD UNIVERSITY PRESS

Providing a template with multiple trapeziums can be helpful for students to draw a line and label the pairs of polygons they make. Provide students with a copy of **BLM 21: Two polygons from a trapezium**. Alternatively, provide students with one trapezium drawn on paper, some thin satay sticks to indicate where they would draw the line, and a recording sheet to complete.

BLM 21

Oxford
OWL

You might provide sentence starters such as: *I can make a _____ and a _____.*

TWO POLYGONS FROM A TRAPEZIUM

Imagine that you draw a straight line through a trapezium.

What are some of the possible pairs of shapes you might make?

What do you notice about the pairs of shapes you have made?

How might you classify the shapes?

Comment

Note that the suggested trapezium is an irregular one. There are many possible pairs, so a key aspect of the task is students recording their thinking. Encourage students to name the different polygons they have made.

Enabling prompt

Give students a paper copy of a regular trapezium and ask:

How can you fold or cut this trapezium to make one or more triangles?

Show me a different way.

Extending prompt

Convince me that you have found all possibilities.

CONSOLIDATING THE LEARNING

There are three additional tasks that require students to consider the ways in which trapeziums can be divided/partitioned into smaller polygons. The intention is that they are able to use appropriate language and justify reasons when discussing their solutions.

WHAT IS THE RULE? 1

What is a rule for cutting a trapezium and getting a pentagon and something else?

Provide examples to support your reasoning.

WHAT IS THE RULE? 2

What is a rule for cutting a trapezium and getting two trapeziums?

WHAT IS THE RULE? 3

Make up your own conjecture about polygons created when you draw a line through a trapezium and ask a partner to prove or disprove it.

SUGGESTION 4: MAKING POLYGONS FROM OTHER QUADRILATERALS

MATHEMATICAL FOCUS

Other quadrilaterals can be decomposed to make polygons.

PEDAGOGICAL FOCUS

The same considerations in the previous suggestion also apply here. In particular, a key component is visualisation, and students explaining and justifying their thinking. Providing students with a thin satay stick can support students when exploring the possibilities of decomposition (dividing or partitioning polygons) before recording their solutions.

Use a similar launch to the previous suggestion, except with the emphasis on parallelograms and their properties, and decomposing (partitioning) parallelograms to create different polygons. You might ask students to discuss what makes a parallelogram different from a trapezium. There may be parallelograms around the school.

The following are prompts you might consider as part of lesson summaries:

- What polygons are produced when the straight-line cut is made on the diagonal?
- What did you need to think about when making a cut to produce two polygons the same, one polygon the same as the original or no polygon the same as the original polygon?
- What is the same and different about the polygons produced?

TWO POLYGONS FROM A PARALLELOGRAM

When you draw one straight line through a parallelogram you get pairs of polygons. Record some possibilities using a sentence like, *I can make a _____ and a _____.*

Enabling prompt

Give the students a paper copy of a parallelogram.

Tell me something about this polygon. If you fold or cut this parallelogram, can you make a triangle? Can you find another triangle?

Extending prompt

When you draw one straight line through a parallelogram do you get more pairs of polygons than if you did the same with a trapezium? Explain your reasoning.

CONSOLIDATING THE LEARNING

There are further tasks that require students to decompose (divide/partition) parallelograms and rectangles by drawing straight lines, or a straight cut to create polygons, and to form generalisations about the properties of parallelograms, rectangles and other quadrilaterals. The emphasis is on students being able to justify their thinking.

OXFORD UNIVERSITY PRESS

POLYGONS FROM A RECTANGLE

What other polygons can you make when you use one straight line to split a rectangle?

Record the possibilities. How do you know you have all the solutions? What are the names of the smaller polygons? Explain how you know.

POLYGONS FROM AN IRREGULAR QUADRILATERAL

What other polygons can you make when you use one straight line to split an irregular quadrilateral?

Record the possibilities. How do you know you have all the solutions? What are the names of the smaller polygons? Explain how you know.

SUGGESTION 5: MAKING POLYGONS FROM OTHER POLYGONS

MATHEMATICAL FOCUS

Polygons can be split and combined to make different polygons.

PEDAGOGICAL FOCUS

A key focus in this suggestion is on decomposing polygons, rearranging the pieces and combining them to make new polygons. As with the previous suggestions, encourage students to visualise the moves first before enacting them.

There are many varieties of tangrams. This suggestion uses a simple version you can create, although feel free to use more complicated versions if your students are ready. These tasks use kinder (kindergarten) squares, although any squares will do.

POLYGONS FROM A KINDER SQUARE ①

Cut a kinder square in half and then cut one of the pieces in half again. How many different ways can you do it?

Comment

The following are the possibilities.

CONSOLIDATING THE LEARNING

There are two additional tasks that challenge students to consider the ways in which polygons can be rearranged to create new polygons.

BARRIER GAME

Using some pieces from your three-piece puzzle, compose a design by placing some or all of the pieces together, but don't let your partner see. Make sure that each piece is connected by at least one side. Place a barrier between yourself and your partner. Give your partner instructions on how to use the pieces to create your design.

POLYGONS FROM A KINDER SQUARE ②

Use the three pieces from one of your possibilities in 'Polygons from a kinder square (1)' to make some new polygons by rearranging and combining the pieces.

What different polygons are possible?

Draw each of these new polygons.

Comment

You might extend this task by encouraging students to look at the possibilities for each of the four ways to cut the kinder square. Below are some possible solutions for each of the four ways to cut the kinder squares (not counting the square).

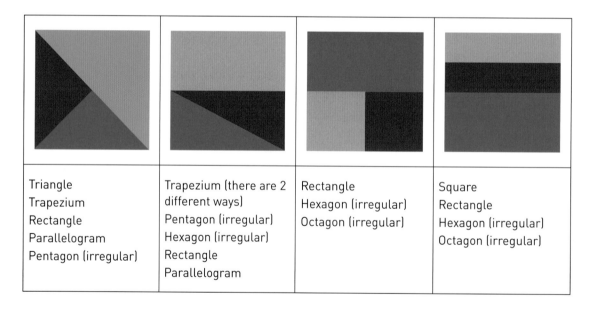

Triangle Trapezium Rectangle Parallelogram Pentagon (irregular)	Trapezium (there are 2 different ways) Pentagon (irregular) Hexagon (irregular) Rectangle Parallelogram	Rectangle Hexagon (irregular) Octagon (irregular)	Square Rectangle Hexagon (irregular) Octagon (irregular)

OXFORD UNIVERSITY PRESS

CHAPTER 13

LOCATION AND TRANSFORMATION

OVERVIEW

The early suggestions in this sequence are suitable for Foundation and Year 1 students, and the latter ones are intended for Year 2 students. There are three related ideas in the sequence: giving and interpreting instructions related to directions; spatial and line symmetry; and rotational symmetry. This sequence is intended to follow on from the 'Recognising polygons' and 'Reasoning with polygons' sequences.

The following is a summary of the suggestions.

	MATHEMATICAL FOCUS
Making maps	Maps are plans that show different pathways to and from a location.
Directions	Directional language is used when giving and following directions.
Exploring symmetry	Line and rotational symmetry are properties of figures and objects.
Flips and turns	Geometric transformations can be created by sliding, reflecting or rotating the position of a figure.

RATIONALE

The focus of this suggestion is on the type of imagining needed when using maps and solving problems related to diagrams and images. The experiences lay the foundation for map reading, angles and transformations, which form a key part of the curriculum in the middle primary years.

The four suggestions focus on different aspects of imagining, especially anticipating the connections between two-dimensional representations of three-dimensional space.

A key feature throughout this sequence is for students to be able to use descriptive spatial language to explain their thinking and or communicate with others.

LANGUAGE

right, left, quarter turn, right angle, flip, reflect, mirror line, symmetry, right angle, quarter turn, imagine, rotational symmetry, line of symmetry, visualise, around, past, over, after, before

LAUNCHING THE TASKS

Many of the concepts in this sequence are new to students. Most of the tasks are posed in the context of problem solving, and so it can help if students are introduced to the language, especially the new language. Some specific ideas for lesson introductions are made within each suggestion.

The transformation sequence is comprised of various aspects, so it is important to consider the concept you intend to assess. For example, *symmetry* (suggested pre-test and post-test assessment task: Suggestion 3: 'Line of symmetry (1)').

The following are some specific statements to inform assessment.
Students are learning about maps, right angles and symmetry when they can:
- give and receive directions (e.g. right, left, forward) depending on the facing direction
- identify and use half and quarter turns (right angles) and solve problems involving right angles
- recognise and create shapes using flips and reflections (line symmetry)
- recognise and create shapes using rotational symmetry (turns).

SUGGESTION 1: MAKING MAPS

MATHEMATICAL FOCUS

Maps are plans that show different pathways to and from a location.

PEDAGOGICAL FOCUS

In this first suggestion, students explore and develop an understanding of drawing and reading simple and familiar maps, and providing directions for routes. When students describe the routes or directions, this encourages them to use a broad range of language including (but not limited to) *around*, *past*, *over*, *near*, *far*, *after*, *before* and *turn*. At this stage the measurements and scale are not important.

The aim of this suggestion is for students to draw and read maps they are familiar with. A key aspect of any of these experiences is the class discussion about the maps and stories that students create. Using the language of location in the context of stories helps to consolidate meaning. You might also have a collection of different maps for the students to explore, such as a map of the school, local area, zoo or botanical gardens.

Links with children's literature, particularly fairy tales, have been made to provide a familiar context. Should your students not be familiar with these stories, consider reading them as a class or changing the context to a familiar story. Examples of other children's books include *Rosie's Walk* (Pat Hutchins); *Knuffle Bunny* (Mo Willems); *We're Going on a Bear Hunt* (Michael Rosen and Helen Oxenbury); and *Piggies in the Pumpkin Patch* (Mary Peterson).

For this suggestion there are no enabling or extending prompts provided. Reviewing student work samples and discussing them can provide both enabling support and ideas to extend students. For students who do need further extension, they could use grids to draw their maps and provide directions. They could also find more than one pathway.

Prior to launching the tasks in this suggestion, we recommend asking the students what they know about fairy tales, particularly *Little Red Riding Hood*. You may like to read a version to the class. The University of Pittsburgh has an online version that is rich with language directions (https://sites.pitt.edu/~dash/type0333.html#perrault) describing the path Little Red Riding Hood and the Big Bad Wolf take.

Discuss the elements that are often found on a map and the similarities and differences between them. Having access to a range of maps for students to view can support them to connect with the elements on a map. Alternatively, this could also be conducted as a part of the summary of the main task, where students could draw upon their experiences of creating maps to identify key elements of a map.

The tasks in this suggestion are suitable for Foundation and Year 1.

LITTLE RED RIDING HOOD

Create a map showing Little Red Riding Hood's journey to grandma's house.

Remember that the wolf takes a shortcut to get there.

Describe the routes they both take.

CONSOLIDATING THE LEARNING

There are further tasks to consolidate the learning from the first one.

THE THREE LITTLE PIGS

Draw a map showing where the Three Little Pigs live. Describe how the third little pig can walk back home so she avoids the path to the Big Bad Wolf.

FARM PATHS

Read *What the Ladybird Heard* (Julia Donaldson and Lydia Monks) as part of the preliminary experience for this task.

Draw a map of a farm that includes at least a front gate and a pond far away. Describe the path to walk from the front gate to the pond.

WALKING HOME

Draw a map showing the path that you take to get home from school. Include as many places of interest that you usually go past as you can think of. This may include a playground, shops, friends' homes, or train/tram/bus stops.

STORY WRITING

Write a short story involving places and ways of getting there and draw a map of the characters' movements.

MATHEMATICAL FOCUS

Directional language is used when giving and following directions.

PEDAGOGICAL FOCUS

This suggestion follows from the suggestions about creating maps. It focuses on right and left turns and moving forward. The intent for students is to learn to follow and give instructions. The challenge for students is to imagine the direction they are heading in, even if that is not the way the diagram is facing.

   The skills involved are similar to those needed to give instructions to robots.

   Many of the tasks assume students are using grid paper.

   The tasks in this suggestion are suitable for Years 1 and 2.

   When launching the 'Walking from A to B (1)' task, we suggest introducing the idea of following instructions by asking students to draw shapes on grid paper, following your instructions. For example, *Put your pen at one corner of a square near the middle of the page. Go up 4 squares along that line. Turn right and go forward 2 squares. Turn right and go forward 2 squares. Describe the shape you have drawn.* (An upside-down U.)

   There may be a grid in the school, such as a large chessboard, that you could use for students to follow the instructions physically. You could create such a grid in the classroom.

WALKING FROM A TO B ➊

The blue lines are footpaths.
Give someone instructions on how to walk from A to B.

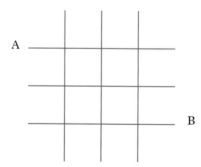

Enabling prompt

While the focus is on students imagining, it may help to have arrow cards that can be used to prompt students experiencing difficulties in awareness of the direction they are facing.

Extending prompt

How many different paths can you make so long as you keep moving towards B?

CONSOLIDATING THE LEARNING

There are further tasks to consolidate the learning from the first one.

WALKING FROM A TO B ❷

The blue lines are footpaths. It is not possible to walk through the blue squares.

Give someone instructions on how to walk from A to B.

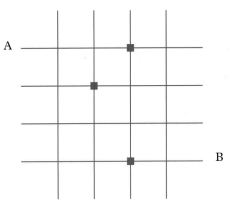

WITHOUT LIFTING YOUR PEN ❶

Write instructions on how to draw this blue shape without lifting your pen and without going along any line twice.

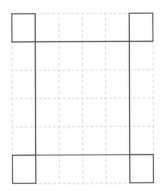

WITHOUT LIFTING YOUR PEN ❷

Write instructions on how to draw this blue shape.

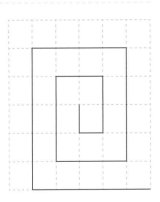

MYSTERY ANT TRAIL

I used the following cards to create the path left by the ants in my house: forward 1, forward 1, forward 1, forward 1, forward 1, forward 1, forward 1, forward 1, turn right, turn right, turn right, turn left, the end!

Map out your ant trail using the cards and then draw what the trail might look like.

MYSTERY ROBOT TRAIL

Suppose you create a walk for a human robot. You have these 12 cards: forward 1, forward 1, forward 1, forward 2, forward 2, forward 3, forward 3, turn left, turn left, turn right, turn right, turn right. What might the robot walk look like?

Give as many possibilities as you can.

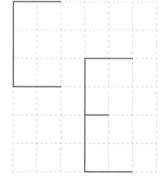

SUGGESTION 3: EXPLORING SYMMETRY

MATHEMATICAL FOCUS

Line and rotational symmetry are properties of figures and objects.

PEDAGOGICAL FOCUS

Young children's experiences of symmetry begin when exploring the world around them, such as in nature, buildings and artwork. Line and rotational symmetry are properties of figures and objects. A shape has line symmetry if it can be divided into two equal parts, one of which is the mirror image of the other. Some shapes have two or more lines of symmetry, whereas figures with no lines of symmetry are asymmetric. A shape has rotational symmetry when it can be turned around a central point less than a full turn, to fit onto itself.

The intention of this suggestion is for students to develop an understanding of line and rotational symmetry through engaging with these tasks, to recognise when shapes have line or rotational symmetry and to understand why others do not. Encourage students to visualise the line of symmetry and then check by drawing it, by cutting out the letters and words and folding them, or by using a mirror. After students have explored the first task, ask them to put their detective hat on and explain what they notice about the letters and words that have line symmetry, compared to those that are asymmetric. Repeat this with the rotational symmetry task. Record their ideas for later reference.

This suggestion is suitable for Year 2 students.

Prior to launching the 'Line of symmetry (1)' task, we suggest providing opportunities for students to explore some symmetrical images of nature, such as butterflies, starfish, leaves, flowers, insects and fruit. What do they notice about them? What is the same? Also include some non-examples as a way to focus on what is different and why they do not belong with the others.

Other examples include:

- 'Copycat': One student moves their body into a shape and their partner then pretends to be the mirror and reflects the same position. This can be done face to face or side by side.
- 'Butterflies line symmetry': View images of butterflies, create your own (collage, playdough or draw), or use the interactive website Weave Silk, http://weavesilk.com.

For the first task, provide students with grid paper to record their solutions. Some of the tasks in the suggestion use block letters. This diagram shows the letters C and E drawn on grid paper.

LINE OF SYMMETRY 1

When written in block capital letters, which letters have a line of symmetry?

What are some words that can be made out of letters with a line of symmetry?

Which of those words have a vertical line of symmetry?

Which of those words have a horizontal line of symmetry?

OXFORD UNIVERSITY PRESS

Comment

Sixteen block letters of the alphabet have a line of symmetry (AMTUVWY – vertical line of symmetry; BCDEK – horizontal line of symmetry; HIXO – both vertical and horizontal line of symmetry).

Enabling prompt

Which of these letters have a line of symmetry?

A B C D E Z

Extending prompt

What is the longest word you can create that has vertical symmetry? What is the longest word you can create that has horizontal symmetry?

CONSOLIDATING THE LEARNING

BLM 22
BLM 23

Oxford OWL

There are five further tasks that could be used to consolidate the learning from the first one. These extend to combine an exploration of line symmetry and rotational symmetry.

Two BLMs are provided to support these tasks: **BLM 22: Reflective symmetry** and **BLM 23: Symmetry sort cards**. The cards in each set need to be cut up in advance.

LINE OF SYMMETRY ❷

Find as many different numbers as you can between 100 and 1000 that have a line of symmetry. What do you notice about the digits in these numbers? What do you notice about these numbers?

SYMMETRICAL DESIGN

Here is a design that is asymmetrical. Redesign it so that it is symmetrical. Draw the new design on isometric paper.

How might you check that it is symmetrical? What actions did you make in recreating your design?

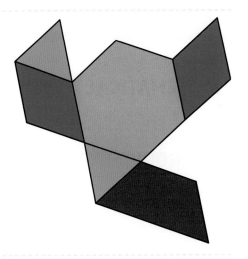

REFLECTIVE SYMMETRY CARD MATCH

BLM 22

Oxford OWL

Provide students with a copy of **BLM 22: Reflective symmetry**. There are two sets of cards. Every card in the red set has a reflected image in the green set.

Find all the reflected pairs.

Enabling prompt

Select 1 card and draw the reflection.

Extending prompt

Choose a card from the red set. Draw what the vertical reflection may look like.

ROTATIONAL SYMMETRY

When written in block letters, which letters have rotational symmetry?

What are some words that can be made out of letters that have rotational symmetry?

What are some words with rotational symmetry?

SYMMETRICAL QUADRILATERALS

Draw some quadrilaterals with one or more lines of symmetry but no rotational symmetry.

Draw some quadrilaterals with rotational symmetry but with no lines of symmetry.

What do you notice about each of these groups of quadrilaterals?

SYMMETRY CARD SORT

BLM 23

Oxford
OWL

Provide students with a copy of **BLM 23: Symmetry sort cards**.

What do you notice about these cards?

Match the pair of images, the action and the type of symmetry.

What other images could you draw that fit the categories? Draw a pair of your own images.

SUGGESTION 4: FLIPS AND TURNS

MATHEMATICAL FOCUS

Geometric transformations can be created by sliding, reflecting or rotating the position of a figure.

PEDAGOGICAL FOCUS

The tasks in this suggestion extend on the idea of symmetry to explore geometric transformations, including changes in the position of a figure through the action of flip (reflect) or turn (rotate). Everyday experiences, such as block play, completing jigsaw puzzles and puzzle games such as Tetris involve these actions and support students' developing mental rotation skills. It is important that students notice that the figure does not change as a result of these actions; it is identical (congruent) – that is, has the same size and the same shape.

As students engage with the tasks, encourage them to imagine what a shape or figure might look like if it is flipped or turned. Ask students to explain the action they used and listen for students using language related to symmetry such as *same, flip, turn, rotate, reflect, mirror, line of symmetry* as well as spatial language such as *same, on top of, below, beside*.

OXFORD UNIVERSITY PRESS

Tasks in this suggestion are suitable for students in Year 2.

Prior to launching the 'Flipping and rotating (1)' task, we suggest providing opportunities for students to build spatial visualisation and imagery and mental rotation skills by doing some short sharp activities such as these:

- 'Quick as a flash': Show students an image of a figure for 2 seconds and then show it again, this time rotated. How is the second figure different?
- Variation: Show a simple composition of two shapes (e.g. a trapezium and a square) for 2 seconds, then ask students to draw what they saw, and describe where they placed the square.

Repeat the activity with different compositions using tangram pieces. Have students construct the composition rather than draw it, and ask them to describe the actions used to construct it.

- *What might it look like now?* Show students a right-angled triangle for 2 seconds, then ask them to imagine they had flipped it and draw what it might look like.

FLIPPING AND ROTATING

On grid paper, write the number '21' as it appears on a calculator (using only right angles).
Flip the number three different ways.
Rotate the number three different ways.

Comment

Encourage students to justify how they know each reflection is a mirror image of the original number and to explain the different ways they rotated the number. What did you need to think about when you flipped the numbers? Some students might not recognise that the number can be flipped left and right as well as vertically, so you might use the enabling prompt.

Enabling prompt

Look at these two figures. The first figure was flipped to get the second figure. How was it flipped? What is another way it might have been flipped?

Extending prompt

Choose a different 2-digit number and flip and then rotate it in 3 different ways. What do you need to think about when you flip a number that you did not think about when you rotated a number? Why?

CONSOLIDATING THE LEARNING

There are further tasks to consolidate the learning from the first one. In each instance, ask students to visualise what the figure might look like after the transformation (flip, slide or turn) and then to describe the transformation that was used. After students have completed the 'Design a mat' and 'Design a patchwork quilt' tasks, ask them to describe and record the turns (e.g. quarter turns, half turns, three-quarter turns) and the reflections (e.g. flip horizontally, flip vertically).

FLIPPING AND ROTATING ❷

The shape on the right has been turned by a quarter turn from the one on the left.

Draw what the shape would look like after one further turn.

Make up two more shapes (like these) where one is a quarter turn of the other.

Make up two more shapes where one is a reflection of the other.

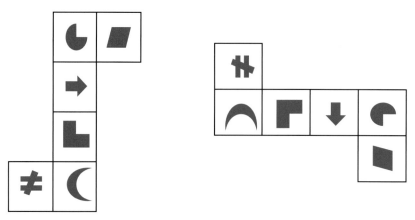

MISSING STEPS ❶

I began with the shape on the left and ended up with the design on the right.

What steps might I have done between the first and the finished design?

Imagine the steps, then draw them and check on a geoboard. What flips and turns did you do?

Make up a different set of flips and turns to get from start to finish.

Start

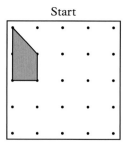

Finish

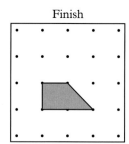

MISSING STEPS ❷

I started with this L shape and finished with the shape in this position. What flips, slides or turns might I have done from the start to the finish?

Create your own design pattern that includes flips, slides and turns.

OXFORD UNIVERSITY PRESS

DESIGN A MAT

This is the first row of a tiled design for the playroom.

There are more rows in the design.

Create more rows by flipping or turning some of these original images.

What might the design look like? What actions did you use to create the design?

What might it look like if you just flipped the images vertically or horizontally?

DESIGN A PATCHWORK QUILT 1

For this task, provide the template shown.

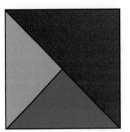

Alternatively, students can create their own template by using these three pieces (the medium triangle and two small triangles from a tangram) to form a square.

This is the first square of a patchwork quilt.

Draw what the other three squares might look like.

Write a list of the turns and flips you used, so a friend can create your design.

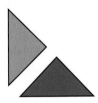

DESIGN A PATCHWORK QUILT 2

This is the square design used to create a patchwork quilt.

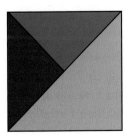

The complete patchwork quilt has 9 squares arranged in a three-by-three array.

What flips, slides or turns might you do to create the quilt?

Draw what the quilt might look like.

How many different patterns can you create?

If you just use quarter turns, half turns and three-quarter turns, what might a quilt look like?

CHAPTER 14

PROBABILITY AND STATISTICS

OVERVIEW

This sequence is suitable for students from Foundation to Year 2. The first three suggestions are suitable for all three year levels, while Suggestions 4 and 5 are more suitable for Year 2.

The following is a summary of the suggestions in the sequence.

	MATHEMATICAL FOCUS
The language of chance events	Particular everyday words describe the likelihood of an event happening.
Asking and answering questions	Statistics is about asking and answering questions about our world. Graphs are used to represent the data.
Using chance words to describe number properties	Chance words can be used to make generalisations about mathematical ideas.
Dice have no memory	Past outcomes of events do not influence future chances.
Data can inform us about chance	Data can be used to learn about how likely something is to happen.

RATIONALE

The advantage of including probability and statistics in early years teaching is that they are experiential, and generally have minimal demand for arithmetic skills but still create opportunities for problem solving and reasoning. There is less emphasis on correct and incorrect answers; instead, the focus is on the meaning of chance words and data words. Like much of mathematics learning, the first step is language. It is hard to solve problems or justify reasoning without the technical words and phrases.

Playing dice games (e.g. snakes and ladders) allows consideration of the chance of rolling a particular number (e.g. the player has a 1 in 6 chance of sliding down the snake, as they need to roll a 2 for this to happen). Students benefit from experiences that lead to the conclusion that 'dice have no memory'.

Likewise, teachers can lead the class in the processes of designing, conducting, recording (especially tallying), representing and inferring as part of this sequence. Foundation level survey questions are mainly answered yes/no. At other year levels, students could have cards on which their names are written and use them to record answers to survey questions such as: the month of their birthday, their favourite sporting team (with a 'no favourite' option) or their favourite food. Part of the focus is on finding different ways of representing the same data.

OXFORD UNIVERSITY PRESS

In both dice games and surveys, it is important to pose questions such as the ones in this sequence.

The tasks in the following suggestions are written with the expectation of students writing their responses, although teachers should choose ways of representing responses that are appropriate to the class. Foundation students can explain their answers verbally. Since the tasks are experiential, having both low floors and high ceilings, there is likely to be no need for enabling and extending prompts.

LANGUAGE

chance, unlikely, likely, certain, possible, impossible, will happen, won't happen, might happen, event, outcome, data, survey, information, tallying

LAUNCHING THE TASKS

You might pose or arrange preliminary experiences that can act as revision or prerequisite skills for each suggestion. Some of the experiences mentioned above (dice games, surveys) could be posed as part of the launch, as appropriate.

You might lead discussions prompted by questions such as:

Who has heard the word certainly?

How many more students have birthdays in June than in July?

Fluency activities focusing on counting by 5s can provide practice for tally counting.

ASSESSMENT

Suggested pre- and post-test assessment task: Suggestion 3: 'What will happen?'

The following are some specific statements to inform assessment.

Students are learning about chance and data when they can:

- use chance words appropriately to describe how likely things are to happen
- use chance words appropriately to make generalisations about number properties
- order chance words based on their meaning of how likely things are to happen
- use data to make inferences about the likelihood of events
- interpret picture and column graphs and make inferences from the graphs
- represent the same data in different ways
- design and conduct surveys, record data, and draw graphs to represent the data
- use the tallying method and say the total of a set of tally marks.

SUGGESTION 1: THE LANGUAGE OF CHANCE EVENTS

MATHEMATICAL FOCUS

Particular everyday words describe the likelihood of an event happening.

PEDAGOGICAL FOCUS

The focus in this suggestion is on hearing and using words that describe the chance that something will happen. These tasks provide opportunities for students to draw a picture, or write or tell a story. Ensure the lesson focuses on the language by encouraging students to verbalise their stories as much as possible. Invite questions, affirm correct usage, and challenge incorrect responses.

The tasks in this suggestion are suitable for Years 1 and 2, or perhaps near the end of the year in Foundation.

CREATE A STORY **1**

Create a story that includes each of the following words at least once:

will   won't   might

CONSOLIDATING THE LEARNING

There are further tasks that consolidate the learning from the 'Create a story' tasks. It is assumed you will create further lessons based on these ideas if needed.

CREATE A STORY **2**

Create a story that includes the following words:

certain   likely   unlikely   impossible

CREATE A STORY **3**

Finish these sentences and create a story.

I will never…

I might…

I will probably…

I will certainly…

CREATE A STORY **4**

Imagine you are playing Snakes and Ladders and you have just landed on 32.

Finish the sentences and create a story.

I will never…

I might…

I will probably…

I will certainly…

Using a chance line, position the following words on the line:

Never

50/50

Unlikely

Likely

Impossible

Might

Definitely

Certain

What other words can you think of that say how likely things are to happen? Justify where they may go on your chance line.

Comment

A chance line operates exactly like a number line, except the position of the word along the line indicates how likely something is to happen. For example, 'never' would be at the far left-hand side of the line, '50/50' would be in the middle, and 'certain' would be at the far right-hand side.

Note that some words (e.g. 'might') will mean different things to students depending on their personal experiences.

SUGGESTION 2: ASKING AND ANSWERING QUESTIONS

MATHEMATICAL FOCUS

Statistics is about asking and answering questions about our world. Graphs are used to represent the data.

PEDAGOGICAL FOCUS

Statistics is more than just creating graphs and analysing data; it is about asking and answering questions about our world. When students are given opportunities to create their own questions, the data is more meaningful to them.

It is assumed that the teacher will lead students in conducting surveys and representing the data. For example, the teacher might suggest to students working in groups: *Make a graph using one shoe from everyone in your group. Explain your graph.*

It is important that students have opportunities to discuss how well-defined their question is.

The tasks in this suggestion involve student choice in their response. Ensure students have opportunities to explain and justify their choices. These explanations should be meaningfully based on the problem posed.

A group of friends made this graph.

How many friends might there have been?

What might have been the survey question they were answering?

What does the picture graph tell us?

Represent the same data in a different way.

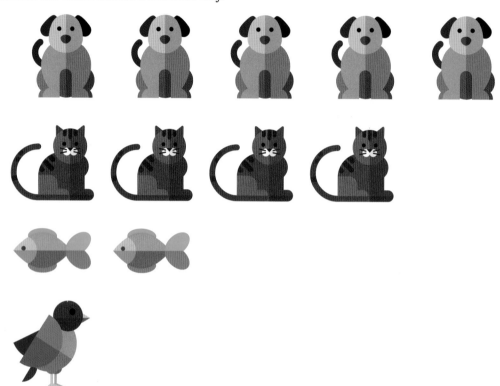

Comment

In the early years, it is best to start with one picture representing one object – in this case, animals. The answer to the question about the number of friends depends on the question students create. If, for example, the question is 'What is your favourite pet?' then there would be 12 friends. If the question is 'What pets do you have?' there can be any number of friends up to 12.

If someone suggests that each picture represents two animals (it is hoped some Year 2 students might suggest this), then the answer is different.

Asking students to create their own graphs shows what other forms of graphs they have seen and how they might interpret the data. A central focus of the suggestion is getting students to show the same data in different ways and to connect different representations (e.g. picture graphs, tally marks, Arabic numerals, graphs).

CONSOLIDATING THE LEARNING

There are further tasks that explore graphs and their representations and consolidate the learning.

SHOW THIS INFORMATION IN A GRAPH

Draw a graph to represent this information.

WHAT WAS THE QUESTION? ②

The teacher did a survey of student opinions and drew this graph. What question were the students answering?

Represent the information another way.

WHAT WAS THE QUESTION? ③

The students in a class did a survey and drew this graph. What question were the students answering?

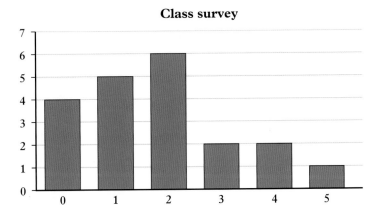

How many students were in the class?

Represent the information another way.

Someone did a survey. What do you think the survey might have been?

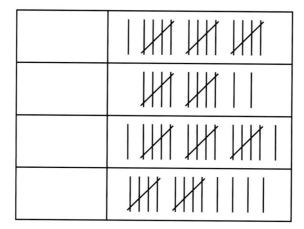

Represent the information another way.

SUGGESTION 3: USING CHANCE WORDS TO DESCRIBE NUMBER PROPERTIES

MATHEMATICAL FOCUS

Chance words can be used to make generalisations about mathematical ideas.

PEDAGOGICAL FOCUS

The focus in these tasks is making generalisations about other aspects of mathematics, mainly number. In each case, discuss the students' answers and encourage them to justify their answers.

WHAT WILL HAPPEN? **1**

If I add an odd number and an even number, what:

a   will never happen?   b   will always happen?   c   might happen?

If I add 10 to a number, what:

a   will never happen?   b   will always happen?   c   might happen?

If I count by fives starting from zero, what:

a   will never happen   b   will always happen?   c   might happen?

Examples of possible responses to the first question might be:

- "I will never get an even number as the answer."
- "I will always get an odd number as the answer."
- "I might get a 2-digit odd number."

Comment

If students find this easy, you might ask them to explore (perhaps using a calculator) what will happen if they multiply an odd number by an even number.

CONSOLIDATING THE LEARNING

There are further tasks that explore chance words with other mathematical ideas.

WHAT WILL HAPPEN? 2

Complete these sentences for your favourite polygon.

A _____ always has _____.

A _____ sometimes has _____.

A _____ never has _____.

WHAT WILL HAPPEN? 3

The minute hand on a clock is on the 6. Complete these sentences.

It is always _____.

It might be _____.

It cannot be _____.

WHAT WILL HAPPEN? 4

Complete these sentences.

If I add 10 to a number that ends in a 2, the answer always …

If I add 10 to a number that ends in a 2, the answer never …

If I add 10 to a number that ends in a 2, the answer might …

WHAT WILL HAPPEN? 5

Complete these sentences.

If the difference between two numbers is 20, the numbers will always be …

If the difference between two numbers is 20, the numbers will never be …

If the difference between two numbers is 20, the numbers will sometimes be …

SUGGESTION 4: DICE HAVE NO MEMORY

MATHEMATICAL FOCUS

Past outcomes of events do not influence future chances.

PEDAGOGICAL FOCUS

This suggestion is recommended for Year 2 students. As with the earlier suggestions, the focus is on students making choices and justifying those choices. The goal is for students to become familiar with the likelihood of the various sums that can be generated from rolling two dice (e.g. rolling a total of seven is more likely than rolling a total of 12).

Students often have considerable difficulty with the notion that dice have no memory, and so past outcomes do not influence the probability for the next roll. This suggestion provides opportunities for students to explore this principle whilst justifying their choices. Whilst it is possible to use physical dice, you may also consider virtual interactive dice.

TWO DICE ❶

When you roll two dice and add the total…
What is an outcome that is impossible?
What is an outcome that is unlikely?
What is an outcome that is likely?
What is an outcome that is certain?

CONSOLIDATING THE LEARNING

There are further tasks that consolidate the learning from the 'Two dice' task.

TWO DICE ❷

When you roll two dice and find the difference…
What is an outcome that is impossible?
What is an outcome that is unlikely?
What is an outcome that is likely?
What is an outcome that is certain?

ONE DICE

Choose your favourite number (on the dice).
If you roll the dice 6 times, predict how many times you expect your favourite number to come up. Now do it.
If you roll the dice 60 times, predict how many times you expect your favourite number to come up. Now do it.

DICE DIFFERENCE

This is a game for 2 players. Each player rolls 2 dice 20 times.
Player 1 (the person with the longer name) gets a point if the difference is 0, 1 or 2.
Player 2 gets a point if the difference is 3, 4 or 5.
Is this game fair?
If you think the rules are unfair, how could you change the rules to make the game fairer?

OXFORD UNIVERSITY PRESS

SUGGESTION 5: DATA CAN INFORM US ABOUT CHANCE

MATHEMATICAL FOCUS

Data can be used to learn about how likely something is to happen.

PEDAGOGICAL FOCUS

This suggestion is recommended for Year 2 students. Again, the focus is on students making choices and justifying those choices. Building on the previous suggestions, this suggestion requires students to consider the connection between data collected and the likelihood of events occurring. Although dice have no memory, data (even from dice) can still inform us about how likely something is to be true or false. Data and probability are often connected in this way in the real world (e.g. scientists collect data to see how likely it is that their theories are true). This suggestion allows students to explore these ideas in an informal way.

There are many potential variants on the ideas presented below. The important consideration is that students are familiar with the context they are exploring in order to support their mathematical reasoning. You could do the same types of tasks in more challenging contexts (e.g. temperatures recorded across five consecutive days – what season are we most likely to be in?). However, the contexts below have been chosen deliberately because they are easy for students to investigate independently (e.g. asking students in other grades how old they are, rolling different dice).

HOW OLD ARE YOU?

I walk into a classroom in your school and ask five students their age.
They reply:

'9' '10' '8' '9' '9'

What is the most likely classroom that I have walked into?
What is the least likely classroom that I have walked into?
Explain your reasoning.

CONSOLIDATING THE LEARNING

There are further tasks that consolidate the learning from the 'How old are you? (1)' task.

HOW OLD ARE YOU? 2

I walk into a classroom in your school and ask five students in the class their age.
They reply:

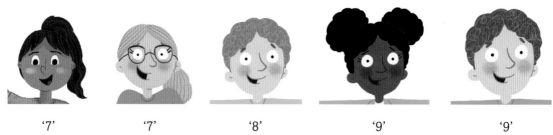

'7' '7' '8' '9' '9'

What is the most likely classroom that I have walked into?
What is the least likely classroom that I have walked into?
Explain your reasoning.

ROLLING DICE 1

I roll a pair of dice and work out the total. I do this 5 times. These are the totals:

7

13

15

12

2

Am I more likely to be rolling a pair of 6-sided dice, 10-sided dice or 20-sided dice? Are all pairs possible? Explain your reasoning.

ROLLING DICE 2

I roll a pair of dice and work out the total. I do this 5 times. These are the totals:

7

11

5

12

2

Am I more likely to be rolling a pair of 6-sided dice, 10-sided dice or 20-sided dice? Are all pairs possible? Explain your reasoning.

OXFORD UNIVERSITY PRESS

CHAPTER 15

APPENDICES

APPENDIX 1: INSTRUCTIONAL MODEL

SEQUENCES OF CONNECTED, CUMULATIVE AND CHALLENGING TASKS FOR TEACHING EARLY YEARS MATHEMATICS

This book presents 14 sequences of learning experiences, arranged according to the key domains of the curriculum, to support the teaching of mathematics in the early school years (students aged 5 to 8). Each sequence is made up of a small number of 'suggestions' that build on one another and support the learning of an aspect of the respective sequence. The following specific information for teachers is embedded within each sequence:

- a summary of the suggestions including a statement of mathematical focus
- an explanation of the rationale for the sequence including big ideas and elaboration of learning goals
- a statement of pre-requisite and new language
- ideas for launching the learning experiences
- specific statements of the intended learning that can inform assessment.

We consider the ideal curriculum to be made up of descriptions of content (nouns, if you like) **and** proficiencies (understanding, fluency, problem solving and reasoning) that can be thought of as verbs. The resources presented in this book are based on the posing of *problems* allowing for student *reasoning* that in turn build relational *understanding* and foster *fluency*. This approach is termed teaching *through* problem solving (Schroeder & Lester, 1989).

The task design and pedagogical emphasis are informed by two characteristics articulated by the Organisation for Economic Co-operation and Development (OECD) (2019; 2021).

The first, *agency*, relates to students having the ability and will to make active decisions to positively influence their own and others' learning. This implies that students see themselves as not only capable of thinking for themselves but also having the confidence and aspiration to do so. To exercise such agency and realise their potential, learners require opportunities and time to explore appropriately challenging problems by themselves initially, prior to explicit teaching. During this time students productively struggle (Sinha & Kapur, 2021) as they make sense of the problem and make decisions about their own strategy and form of representation. Subsequently, students work in pairs or small groups to collaborate on possible solutions.

In terms of emphasising student agency, teachers are encouraged to plan experiences that are appropriately challenging for students. Sullivan et al. (2020) explained that:

> Challenge comes when students do not know how to solve the task and work on the task prior to teacher instruction. Other characteristics of such tasks are that they: build on what students already know; take time; are engaging for students in that they are interested in, and see value in persisting with a task; focus on important aspects of mathematics (hopefully as identified or implied in relevant curriculum documents); are simply posed using a relatable narrative; foster connections within mathematics and across domains ... (pp. 32–33)

The second characteristic, *inclusion*, involves identifying learning experiences and incorporating pedagogies that maximise opportunities for all learners. As far as possible, in the approach proposed in this book, all learners are given opportunities to think for themselves and to access the full curriculum. This connects to experiences in which the activities and tasks are accessible for learners in classrooms while still being productively challenging, and with explicit teacher attention to actions that address needs of individual learners.

There are three aspects of the recommended pedagogies that foster inclusion. First, teachers are encouraged to choose learning experiences that are not only readily accessible for students but also have the potential for further exploration. Second, teachers prepare specific enabling prompts for students experiencing difficulty and extending prompts for students who complete the set work quickly (see Sullivan et al., 2006). Third, teachers consistently use a particular lesson structure that provides students with confidence of what is to come.

We describe approaches to instruction that foster agency and inclusion as *Student-Centred Structured Inquiry*.

Teaching *through* problem solving has also been associated with *cognitive activation*. Compelling evidence of the effectiveness of cognitive activation strategies is reported by Caro et al. (2016) from their analysis of PISA 2012 results involving over 500,000 students. They found that the more often teachers used cognitive activation strategies the more mathematics the students learned.

An instructional model, presented in Figure 3, communicates various associated teacher actions of Student-Centred Structured Inquiry approaches. The language of the instructional model draws heavily on Smith and Stein's (2011) work, which focuses on orchestrating classroom discussions as an essential element of creating opportunities for student agency and ensuring inclusion in the role of the students in creating new knowledge is obvious to them.

Figure 3　Student-centred structured inquiry instructional model

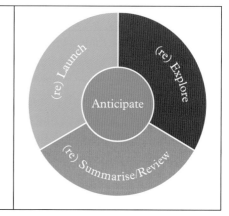

The *Anticipate* phase is, of course, central to all planning. It includes identifying the intended learning and helpful resources, predicting students' solutions, strategies and possible pre/misconceptions, considering pre-requisite and new language, as well as other aspects of planning.

The *Launch* phase includes addressing language and representations associated with the intended learning and practice at fluency relevant for the experience. It also involves posing tasks without informing students on how to solve the problem, an essential aspect of preserving the challenge and fostering student agency.

During the *Explore* phase, teachers interact with students, encouraging persistence, posing prompts to support individual student thinking, and identifying interesting and perhaps unanticipated student responses, selecting some for later presentation.

At the *Summarise/Review* phase, teachers sequence student responses to be shared as a class and support those students while they present their solutions and strategies for subsequent discussion. Further questioning is used to engage students and hold them accountable for communicating their ideas and thinking deeply about the mathematics. A key element of this phase is the teacher synthesising key ideas that represent the learning intentions of the experience by asking questions that focus on mathematical meaning and relationships between the ideas and representations. At this phase, the teacher is explicit about what students are learning, why they are learning it, and how they are intended to learn.

Importantly, the Launch–Explore–Summarise/Review process happens more than once for each learning experience. Each suggestion includes tasks that are 'a bit the same and a bit different' from the initial task. The explicit intention is to *consolidate* thinking activated by the initial experience. This consolidation involves repeating the previous three phases (Launch–Explore–Summarise/Review), noting that consolidation can be in a subsequent lesson.

The sequences are intended to make the central mathematical ideas obvious to the students. Providing opportunities for students to seek insights for themselves along with a consistent lesson structure can reduce the sense of risk that some students experience even in the early years (see Buckley & Sullivan, 2021, for discussion of ways that the instructional model and planned sequences can normalise uncertainty and reduce anxiety).

Even though the mathematical thinking of individual students is paramount, both the intended mathematics and the pedagogical approach are intentional and go beyond unstructured discovery, inquiry or play. At the same time, the approach outlined clearly challenges the assumption that the optimal way to teach mathematics is by explicitly telling students what to do followed by practice. In the implementation of the resources in this book, explicit teaching occurs **after** the learning *through* problem solving in which students have had opportunities to engage with the problem or task prior to hearing explanations of the central ideas from the teacher or other students.

In the research project, within which the resources were developed, we found that teachers who utilised these resources in their classrooms reported improved student engagement, better developed understanding, and more productive dispositions. From a teacher perspective, employing these pedagogies led to greater valuing of productive struggle (Russo et al., 2020), supported differentiated instruction, and fostered a growth mindset characterised by the belief that all students can learn mathematics (Bobis et al., 2021). Teachers also reported that planning time was simplified and reduced since the resources assume whole class mixed achievement student grouping.

We encourage teachers to read fully any sequences they are intending to teach and to imagine the trajectory of learning encapsulated by the resources. The pedagogies have all been trialled in classrooms and have the potential to enhance planning, teaching and assessment of mathematics.

REFERENCES

Bobis, J., Russo, J., Downton, A., Feng, M., Livy, S., McCormick, M., & Sullivan, P. (2021). Instructional moves that increase chances of engaging all students in learning mathematics. *Mathematics, 9*(6), 582.

Buckley, S., & Sullivan, P. (2021). Reframing anxiety and uncertainty in the mathematics classroom. *Mathematics Education Research Journal*, 1–14.

Caro, D., Lenkeit, J., & Kyriades, L. (2016). Teaching strategies and differential effectiveness across learning contexts: Evidence from PISA 2021. *Studies in Education Evaluation, 49*, 30–41.

Organisation for Economic Co-operation Development (OECD) (2019). Future of education and skills 2030: Conceptual learning framework. Concept note: Student agency for 2030. OECD Publishing.

Organisation for Economic Co-operation Development (OECD) (2021). Adapting curriculum to bridge equity gaps: Towards an inclusive curriculum. OECD Publishing.

Russo, J., Bobis, J., Downton, A., Hughes, S., Livy, S., McCormick, M., & Sullivan, P. (2020). Elementary teachers' beliefs on the role of struggle in the mathematics classroom. *The Journal of Mathematical Behavior, 58*, 100774.

Schroeder, T. L., & Lester, F. K. (1989). Developing understanding in mathematics via problem solving. In P. Trafton (Ed.), *New Directions for Elementary School Mathematics* (pp. 31–42). National Council of Teachers of Mathematics.

Sinha, T. & Kapur, M. (2021). When problem solving followed by instruction works: Evidence for productive failure. *Review of Educational Research, 91* (5) 761–798.

Smith, M. S., & Stein, M. K. (2011). *5 practices for orchestrating productive mathematical discussions*. National Council of Teachers of Mathematics.

Sullivan, P., Bobis, J., Downton, A., Hughes, S., Livy, S., McCormick, M., & Russo, J. (2020). Ways that relentless consistency and task variation contribute to teacher and student mathematics learning. In A. Coles (Ed.), *For the Learning of Mathematics Monograph 1: Proceedings of a Symposium on Learning in Honour of Laurinda Brown* (pp. 32–37). FLM Publishing Association.

Sullivan, P., Mousley, J., & Zevenbergen, R. (2006). Developing guidelines for teachers helping students experiencing difficulty in learning mathematics. In P. Grootenboer, R. Zevenbergen & M. Chinnappan (Eds.), *Identities, Cultures and Learning Spaces*. Proceedings of the 29th annual conference of the Mathematics Education Research Group of Australasia. (pp. 496–503). MERGA.

'WHAT ARE SOME OF THE OBSTACLES TO TEACHING WITH SEQUENCES OF CHALLENGING TASKS IN THE FIRST YEARS OF SCHOOLING AND HOW MIGHT TEACHERS OVERCOME THEM?'[1]

Working on challenging tasks requires students to engage in sophisticated mathematical behaviour including: planning how they will approach the task; choosing appropriate strategies and explaining these choices; recording their thinking and using this record to support mathematical reasoning. Equally important are the dispositions that students need to develop when engaged with such tasks. A desirable disposition includes taking personal responsibility for their learning through deciding when they need support (enabling prompts) and when they are ready to take their thinking further (extending prompts). Importantly, it also includes a willingness to tolerate, and even embrace, confusion and to persevere despite experiencing frustration and struggle.

Given the mathematical and psychological demands working on challenging tasks places on students, it is perhaps not surprising that teachers in the early years of schooling, particularly Foundation teachers, are often uncertain about whether this type of mathematical work is appropriate for their students. In this section, we present obstacles articulated by some of the teachers involved in our project when teaching with the sequences of tasks presented in this book, and how these teachers overcame these obstacles. Quotes from project teachers are included throughout this section, both to frame these obstacles and to demonstrate how teachers overcame them.

The four obstacles we discuss are:

1 the belief that challenging tasks are not for all students

2 not knowing when to begin teaching with challenging tasks

3 the feeling that some tasks are not meaningful for students (i.e. are too mathematically abstract)

4 the belief that students cannot cope with the struggle.

OBSTACLE 1: CHALLENGING TASKS ARE NOT FOR ALL STUDENTS

Should these tasks only be for the high flyers?

They were in that low group … I suppose I never wanted to pose the challenge because I didn't want them to think that maths was hard.

---

[1] This appendix draws substantially on the following article: Russo, J., Bobis, J., Downton, A., Hughes, S., Livy, S., McCormick, M., & Sullivan, P. (2019). Teaching with challenging tasks in the first years of school: What are the obstacles and how can teachers overcome them?. *Australian Primary Mathematics Classroom*, *24*(1), pp. 11–18.

OVERCOMING THIS OBSTACLE

When some project teachers first began to use the sequences of challenging tasks in their classrooms, they assumed that because these tasks are demanding, they should be reserved for students identified as being highly mathematically capable. Many teachers overcame this obstacle through changing their own mindset, reconsidering what it means to make progress with a task, and sometimes even reconsidering what it means for a student to be considered mathematically capable.

Several teachers made the point that there can be benefits to exposing students to tasks that would seem unequivocally beyond their mathematical knowledge, because such experiences provide students with opportunities to informally explore sophisticated mathematical ideas, and become familiar with important mathematical tools.

> It's definitely changed my mindset. I exposed all students to a lot more than what I used to. I used to have a mindset that that's a lower cohort, there's no point in doing that with them … now I think "Why not? Why wouldn't I do it?" Because there are some kids that pick things up that you don't realise.
>
> I've been surprised … by doing these tasks that, even if it just gives them an understanding, oh that's a 100s chart, at least they know what it is and things like that … So that has been certainly an eye-opener, to be able to give everyone I suppose the challenge and then just see where they go.

Other teachers noted that working through these sequences revealed that making assumptions about student ability can oftentimes be problematic, as it was not always those categorised as 'top students' who had the most success with these tasks. Engaging with these sequences sometimes revealed that so-called 'top students' previous exposure to mathematics had been narrow and procedurally focused.

> When we started it was very interesting to see how many of my top students really struggled with it. I think because a lot of it is rote learning that they've done at home before they've started school and they might be really good at recognising numbers and doing simple problems and that sort of thing. But they didn't have those skills to think outside of the box and to do some problem solving. It was those middle kids that when they were given those opportunities to have a go at something that was a bit more challenging, that they were the ones that really shone, and I thought that was really interesting.
>
> High achievers initially found this sequence very difficult because of their past experiences of maths being easy … the challenge of new learning was something they had not yet experienced. Many of these students were supported by students that were used to the struggle.

Conversely, it was also noted that 'vulnerable students' or those 'lacking confidence' would frequently surprise teachers with the progress they would make with tasks in the sequences:

> When first implementing the sequences, I would launch students into a task and assume that my 'vulnerable' students wouldn't be able to access the learning. I quickly learnt not to make those assumptions. I was taken aback at the level of work they were producing.
>
> Lilly lacked confidence in Maths and had gaps in her knowledge … Her persistence and interest in listening to other strategies and (to) have a go surprised me. She ended up having some of the most efficient strategies … and was able to explain her thinking proudly.
>
> One of my students at the beginning of doing the sequences would not complete or attempt any task independently without teacher support. He always would come up to me and say

OXFORD UNIVERSITY PRESS

that it was too hard. As we progressed through the sequences he became more independent and gained confidence in his ability to complete tasks. By the end he was able to write, draw and attempt to write a statement about his thinking. The child is now able to explain his thinking to the class and use a variety of strategies to solve problems.

OBSTACLE 2: WHEN TO BEGIN TEACHING WITH CHALLENGING TASKS

We felt a bit hesitant because we felt our students weren't ready.

Initially we got all our preps in, did the early years testing and found out some of them just had no understanding [of numbers] whatsoever.

OVERCOMING THIS OBSTACLE

Many of our project teachers, particularly Foundation teachers, were unsure as to when the most appropriate time was to begin teaching with sequences of challenging tasks. Initially, many waited until the second half of the year to begin exploring the sequences with their students. However, many teachers subsequently expressed the view that waiting until students are suitably settled into the year, and have developed the requisite mathematical knowledge, is unnecessary. As noted previously, such teachers acknowledged that students can explore fundamental mathematical ideas through the process of being exposed to challenging tasks, even if students do not necessarily complete the task as designed. Teachers also described how tasks can be revisited throughout the school year, with different connections and emphases made on each occasion.

> Our advice is don't wait. We've done some of the tasks in Term 4, some of them in Term 3 and in previous years some in Term 1 and we've found something different at each point. So, what our teachers have discovered is some of these tasks can be repeated over and over and over again … Children are in prep and they need multiple exposures of the same concepts so don't think that these sequences and suggestions should just be experienced once, experience them each term, particularly those number ones.

We acknowledge that it is possible that you will remain unconvinced that some students in your learning community early in their first year of school are ready for the experiences offered in the sequences. Consequently, in this book, you will find some suggestions for tasks that you might explore prior to beginning the sequences, particularly for those foundation or kindergarten teachers who feel that the gaps in their student knowledge are too substantial to begin teaching with the sequences as presented.

From discussing this issue at length with project teachers, however, a reluctance to begin using challenging tasks in a classroom may reflect more than concerns about student readiness; it may also imply that teachers have concerns about whether they are ready to teach this way. One of the most powerful comments made during our focus group discussions with project teachers related to the opportunity to reframe one's own anxiety about teaching with sequences of challenging tasks. If a teacher finds the prospect of using these tasks daunting, they can model 'giving it a go', demonstrating how mistakes can be viewed as learning opportunities rather than failure.

In response to the provocation: 'When is the right time to begin teaching with challenging tasks?' a teacher responded, with much affirmation from her peers:

> You jump in and you give it a go … You just do it … because if it doesn't work, that's okay, they're still learning. If you think 'oh I could've done that better' you can do it again and you just keep going with it. Just as it's okay for them to make mistakes it's okay for us to make a mistake and to acknowledge that 'you know what, I didn't do that overly crash hot, I'm going to give that another go' … I think as teachers you can't be afraid.

OBSTACLE 3: SOME TASKS ARE NOT MEANINGFUL FOR STUDENTS

One of the challenges was the minimal prior knowledge that the students are coming in with.

Some tasks weren't relatable for some students.

OVERCOMING THIS OBSTACLE

In our experience, running demonstration lessons in primary classrooms using these sequences of challenging tasks, we have generally found students to be highly interested in the problems posed and in participating in the subsequent mathematical discussion. However, several of our early years project teachers, and Foundation teachers in particular, noted that some of the tasks as they were originally presented can fail to adequately engage all students' attention. This was often attributed to students lacking a meaningful context through which to make sense of the task. According to these teachers, overcoming this obstacle involved finding a way to bring the mathematics to life for the students through identifying a narrative hook, and/or personalising the learning experience.

Research suggests that stories can support the development of mathematical understanding through, for example, emotional engagement, promoting visualisation and mental imagery, and providing shared context for students. All these advantages of using stories in mathematics were in fact raised by our project teachers. In some instances, teachers noted that they launched a challenging task through first reading a picture storybook.

> Just thinking about prep children, when we were getting into the length sequence, talking about prior knowledge and context, we found that it was really helpful to read a story and then no matter what prior knowledge the kids had they all had something to talk about and something to go from. So, we read the King's Foot and then they went off and did activities based on that story which really helped them.

In other cases, teachers reframed the task in a story context to make it more compelling, often using characters known to the students, such as other teachers at the school.

> Sometimes the teachers would just make up make believe stories, especially with the cupcakes, Mr Xxx and Mr Yyy came to school crying this morning because he had one less cupcake, and the children really wanted to help even up the plates … We're working with five year olds, they love that kind of stuff.

Another means of deepening student engagement in the task is to further personalise the learning. Project teachers highlighted how this might involve including students in the actual wording of problems (i.e. as part of the narrative hook).

> They connect to it a lot more if their name is part of the challenge and it's about something that they enjoy … It makes it more meaningful to them.

On occasion, personalising also involved labelling strategies and solution approaches with student names to make them more salient.

> The kids will have explained the split strategy but they won't remember it's called the split strategy, they would remember that it's called Eugene's strategy. So that shows the power I guess in learning from other children.

As noted in the 'How to use this book' section, we have generally left the responsibility of launching the tasks in a meaningful context to teachers. Teachers develop a deep knowledge of their students and school communities. They are better placed than anyone to know the sorts of narrative hooks and efforts at personalisation that are likely to resonate in their classrooms.

OBSTACLE 4: STUDENTS CANNOT COPE WITH THE STRUGGLE

The first time I attempted a challenging task with my prep children, they did not cope well with the 'struggling' and 'challenging' aspect … they didn't like that I wasn't spoon feeding them.

OVERCOMING THIS OBSTACLE

Although generally teachers embraced students being in the 'zone of confusion' (sometimes referred to as the 'learning pit') when they initially explored the task, and viewed this as an important aspect of the teaching approach, for some teachers in our project, accepting student struggle remained a concern. Our advice to teachers confronting this issue is that, if you are willing to persist teaching with challenging tasks, you and your students will get there. In our experience, the evolution of a growth mindset in students is contingent on a teacher initially tolerating, and then normalising, student struggle as an integral aspect of learning. It is such teacher actions – and often lack of action – that begin to sow the seeds of a classroom culture punctuated by curiosity, problem solving and perseverance. This sentiment was appropriately captured by one project teacher:

> At the beginning, they were quite hesitant. But the more we worked through tasks and focused on a growth mindset, the more they became comfortable to take risks and problem solve.

As a rule of thumb, you might work towards providing students with five minutes of independent thinking time when the challenging task is first launched, before you support students with enabling prompts and/or allow them to collaborate with others. This brief independent thinking time serves to make being in the 'zone of confusion' part of the learning routine, whilst also reassuring students that they can access support after a few minutes if required or desired.

CONCLUDING REMARKS

The key unifying mechanism that supported project teachers in overcoming these obstacles was a change in their mindset. It is our view that this change in mindset was primarily driven by a commitment from participating schools that study teachers would actually trial at least some of these sequences of challenging tasks in their classrooms; that is, changes in mindset followed from changes in teacher behaviour. Specifically, as teachers began to use these sequences, they overcame their initial paralysis, that is, 'not knowing when to begin teaching with challenging tasks' and the belief that 'students cannot cope with the struggle'. Moreover, the fact that most students seemed to benefit from exposure to these sequences and tasks helped to undermine the 'belief that challenging tasks are not for all students'. Finally, as these initial pedagogical experiences allowed teachers to gain further confidence, they became more comfortable modifying and adapting tasks, helping them to directly address their concerns 'that some tasks are not meaningful for students'.

Summary of obstacles and how they can be overcome

OBSTACLE	OVERCOMING THE OBSTACLE
The belief that challenging tasks are not for all students	There are benefits to exposing students to tasks and informally exploring ideas. Making assumptions about student ability can be problematic.
Not knowing when to begin teaching with challenging tasks	Tasks can be revisited throughout the school year. Mistakes by students and teachers can be viewed as learning opportunities.
The feeling that some tasks are not meaningful for students	Identify a narrative hook. Personalise the learning.
Students cannot cope with the struggle	Learn to embrace student struggle when students are first exploring a task. Normalise the 'zone of confusion' by building to 5 minutes of independent work time when a task is launched.

OXFORD UNIVERSITY PRESS

THE VERY EARLY LEARNING OF NUMBER[1]

Posing challenging tasks in which students engage prior to instruction when supported by appropriate pedagogies, is effective at engaging most students in creating and learning mathematics. To some extent, most of the resources presented in this section of the book assume that students are already fluent with saying the number sequence forward to 10 or beyond and can allocate the correct number to quantify a collection. Yet some teachers who participated in our project reported that some students required additional support in achieving both of those goals.

In exploring this further, we examined some data from the *Early Numeracy Research Project* (ENRP) (Clarke et al., 2002) which involved 868 one-on-one interviews with trained interviewers using a structured format and supplied materials. The breadth of schools participating in the ENRP project provides some confidence that the results are representative of the cohort. The reported results indicated that 35 per cent of the children at the start of school could not say the number names to 20, and 49 per cent could not count a collection of around 20 teddies. Twenty per cent of students could not recognise 2, 0, and 3 dots without counting. In other words, there is a substantial minority of students who had not reached the assumed threshold of the current resources we are trialling as part of our project.

Next are some teaching suggestions for teachers who might have such students in their classes. Note that the intention is to identify experiences in which the whole class participates even though the majority of the class may be fluent with saying the number sequence and counting collections. This is important because it would be most unproductive to separate students experiencing difficulties into small groups and give them instruction different from the rest of the class (see also Clarke, 2021). Not only would this mean that the isolated students would miss some critical new learning opportunities but also such action would communicate to those students that in some way they are different from the others. Inclusion is one of our highest priorities for all children when learning mathematics.

DEVELOPING FAMILIARITY WITH THE NUMBER SEQUENCE (STABLE ORDER PRINCIPLE)

One of the pre-requisites to any interpretation and use of numbers is being able to say the words in sequence and knowing, for example, that the word 'five' comes after the word 'four'. In general, it seems experiences that involve two or more sensory experiences (e.g. sight, sound or touch) can assist students in becoming familiar with the sequence of the words while still being engaging for other students who know the sequence. Information can be learned and remembered better when presented across two experiences rather than one. This approach to learning is also consistent with cognitive science research suggesting that multisensory experiences benefit early number learning when compared with unisensory experiences (Crollen et al., 2020; Jordan &

---

[1] Sullivan, P., Corovic, E., Russo, J. & Hughes, S. (2022). Some steps in the very early learning of number. *Prime Number* *37*(1), pp. 4–8.

Baker, 2011). Our suggestion is to use materials and start by saying the sequence forwards from zero, first up to 10 and later to 20, prior to saying the sequence backwards and going forward starting away from 0.

RHYMES, SONGS AND CHANTS

Rhymes, songs and chants accompanied by some kind of movement or action that link to the sequence are helpful. There are many useful rhymes and songs. An example is

One little, two little, three little friends, etc. up to

Ten little lovely friends.

with the fingers displayed to represent the friends. It can help to use different versions of the same rhyme. For example, the rhyme might be

One jump, two jumps, three jumps of kangaroos, etc. up to

Ten jumps of kan-gar-oos

with students jumping along the basketball court. Once children become familiar with such rhymes and the accompanying finger actions, the teacher may choose to thread in an additional representation – Arabic numerals (digits 0 to 9) – to further strengthen children's familiarity with the number sequence. For example, as a small group activity, the teacher may choose to draw a number line in chalk with the numbers 0 to 10 (and later 0 to 20), with the children jumping along the number line as they sing the kangaroo rhyme. Again, there is evidence from the cognitive sciences that such 'embodied' experiences with number lines can be effective in enhancing children's number sense (Fischer et al., 2011).

Chorusing (reciting the rhyme as a group) is powerful in that it can be self-correcting and students are at least hearing the sequence even if not saying it.

COUNTING WITH PHYSICAL ACTIONS

Similar, but without the group chorusing, is counting forward and back and linking it with a movement such as standing up/sitting down, jumping jacks or turning around once or clapping.

An activity for small groups can be saying the number sequence forwards from 0. Students stand or sit in a circle with individual students, in turn, indicating who says the next number by pointing to the direction of left or right. The advantage of this game is that students listen to all of the numbers, not just the one from the student who says the number before them.

NUMERAL LINE-UP

Integrating the idea of a number line with a sequence can help to provide a visual representation of number order. It has been suggested that the use of linear representations when playing board games can help improve the numerical knowledge of disadvantaged students (Siegler & Ramani, 2008). With this in mind, another activity for small groups of students is to provide a set of numeral cards for a chosen sequence of numbers. The students each take a card to form a human number line. Next, each student says their numeral aloud to check the number sequence is correct.

CHILDREN'S LITERATURE

Children's literature can be helpful for both counting objects and reinforcing the number sequence. For example, *Pete the Cat and His Four Groovy Buttons* (Litwin, 2012) tells the story of a cat who keeps losing buttons from his shirt. The book emphasises the backward number sequence and conveys this through a song as each button falls from Pete's shirt.

USING NUMBERS TO QUANTIFY A COLLECTION (CARDINALITY)

Once the number sequence is known, the next step is to connect each number word with the number of objects in a collection. There are some advantages in focusing on one number at a time in establishing cardinality (the number of items in a given set), as it can encourage children to make generalisations and deepen their representation of a given quantity. It may also support connections between counting by ones to ascertain a quantity (i.e. the cardinality principle) and attending to structure to ascertain a quantity (e.g. conceptual subitising and knowledge of part–part–whole relations; Kullberg & Björklund, 2020). The following suggestions focus on the number 4. Of course, they should be repeated using different focus numbers.

GROUPS OF 4

Teacher calls an instruction, for example, 'lines of 4' and students physically form into the lines. Students explain their grouping by using the sequence (e.g. 1, 2, 3, 4), or through subitising and drawing on their knowledge of part–whole relationships (e.g. 'I know 2 and 2 more is 4'). The teacher uses the sequence to check the groups.

WHAT IS THERE 4 OF?

The teacher asks the question 'What is there in the room that there are exactly 4 of?' The advantage of an activity like this is that there are many possibilities and students can find their own unique collection. Students show their collection to a partner and use the number sequence, or their understanding of part–part–whole relations, to show that their collection has the right quantity of objects. Again, the teacher uses the number sequence to check the collection.

CHILDREN'S LITERATURE

There are many possible mathematical challenges arising from storybooks. For example, after reading *The Very Blue Thingamajig* (Oliver, 2008) the teacher might ask 'Draw a Number 4 thingamajig that has 4 eyes, 4 ears and 4 legs.'

HIDDEN NUMBERS

Students use classroom art materials and glue the quantity of each item on the plate such as 4 feathers, 4 stickers, and 4 buttons.

CREATING DOT PATTERNS

Show the students a dot pattern for a few seconds, then ask them to re-create the pattern using counters. Discuss how they created the pattern and the number of dots in the whole pattern.

TENZI

In the game Tenzi, each player rolls 10 six-sided dice. With a target of 4, they choose the dice that show 4 dots, then places these dice aside and rolls the remaining dice again. They keep rolling until all dice show 4 dots.

COUNTING ONCE AND ONLY ONCE (ONE-TO-ONE CORRESPONDENCE)

While it might seem obvious that each object in a collection should be counted once only, it is less obvious that a system is needed to ensure this happens. To do this, students benefit from identifying a place to start, a place to finish, and a strategy for including all the objects just once. Note that in many instances, objects are counted without touching them.

MATCHING ONE TO ONE

Students are shown a set of eggs and a set of egg cups, for example, and asked to estimate, 'Can you tell me which has more, the eggs or the egg cups, or are they the same?' Students are asked to show or explain their strategy for matching the eggs and cups. Comparing quantities of objects using mathematical language such as *more*, *less* and the *same* is an important step in helping students to develop concepts of number and counting (Clements & Samara, 2014).

Comparing a collection of water bottles and lids provides an alternative context for a similar learning experience, for example, students could be asked, 'Do we need more lids for these bottles or more bottles for these lids or are they the same amount of each?'

RANDOM COLLECTIONS

Students are shown a randomly placed collection of, say, two different coloured counters. They can be asked to estimate the number of blue counters, then to describe their strategy to count the blues, and then to count them. This can be repeated for the red counters and then the collection overall. Asking students to use numerals to record the count should support them with making connections between the collection of objects, number words and symbols.

CIRCLES AND LINES

Students are shown a set of counters arranged in a circle and a similar collection arranged in a line, students are asked to estimate which collection has more, explain their strategies for counting, and then count each collection to check.

THE QUANTITY IS INDEPENDENT OF THE WAY THE OBJECTS ARE ARRANGED (CONSERVATION)

A common early misconception is that simply moving counters, without adding or removing any, can change the quantity. In fact, this misconception is surprisingly resilient, even into middle primary years.

WHICH ONE IS 4?

Students are shown, for example, 3, 4, and 5 represented on ten-frames in various ways, and asked to identify which ones are 4s.

They can also be asked to make 4 in as many different ways as they can on ten-frames or by using fingers.

MAKE 4

The teacher says, 'Show me 4 fingers. Now, show me 4 fingers a different way.' The use of fingers in representing number quantities is recognised as an important support for developing the concept of counting (Gracia-Bafalluy & Noël, 2008). Such interactions also assist in building fluency with a given quantity (e.g. 'fourness') and support the exploration of part–part–whole relations (Kullberg & Björklund, 2020).

COUNTERS IN A CUP

The teacher drops a counter in a cup and students call out the number of counters in the cup, in this case, 'one'. The teacher drops another counter in the cup and students call out 'two' and so on. On four, for example, the teacher does the same hand motion but does not put a counter in the cup. Alternatively, the teacher shakes the cup and asks children to show with their fingers how many counters are in the cup.

CONCLUSION

The suggestions above are intended to support the learning of students who may experience difficulty in connecting to learning the Australian Curriculum content near the start of their Foundation year. Given that much learning is social, it is argued that the experiences are intended for whole class mixed achievement teaching. We have noticed that experiences involving two or more senses can assist students in learning early number concepts, while still being engaging for those students familiar with counting principles.

By providing students with a range of learning experiences that focus on important ideas in early counting, in particular, number sequences, using numbers to quantify a collection, counting using one-to-one correspondence and number conversation, they will have opportunities to develop their conceptual understanding in each of these areas with the aim of later participating in learning mathematics through challenging tasks without requiring significant enabling prompts or supports.

REFERENCES

Clarke, B., Clarke, D. M., Cheeseman, J., Gervasoni, A., Gronn, D., Horne, M., McDonough, A., Montgomery, P., Rowley, G. & Sullivan, P. (2002). Early numeracy research project: Final report. Melbourne, Victoria, Australia: Department of Education, Employment and Training, the Catholic Education Office, and the Association of Independent Schools.

Clarke, D. (2021). Calling a spade a spade: The impact of within-class ability grouping on opportunity to learn mathematics in the primary school. *Australian Primary Mathematics Classroom, 26*(1), 3–8.

Clements, D. H., & Samara, J. (2014). Learning and teaching early math: The learning trajectories approach. Routledge.

Crollen, V., Noël, M. P., Honoré, N., Degroote, V., & Collignon, O. (2020). Investigating the respective contribution of sensory modalities and spatial disposition in numerical training. *Journal of Experimental Child Psychology, 190*, 104729.

Fischer, U., Moeller, K., Bientzle, M., Cress, U., & Nuerk, H. C. (2011). Sensori-motor spatial training of number magnitude representation. *Psychonomic Bulletin & Review, 18*(1), 177–183.

Gracia-Bafulluy, M., & Noël, M.P. (2008). Does finger training increase young children's numerical performance? *Cortex, 44*(4), 368–375.

Jordan, K. E., & Baker, J. (2011). Multisensory information boosts numerical matching abilities in young children. *Developmental Science, 14*(2), 205–213.

Kullberg, A., & Björklund, C. (2020). Preschoolers' different ways of structuring part-part-whole relations with finger patterns when solving an arithmetic task. *ZDM Mathematics Education.* https://doi.org/10.1007/s11858-019-01119-8.

Litwin, E. (2012). *Pete the Cat and His Four Groovy Buttons.* HarperCollins.

Oliver, N. (2008). *The Very Blue Thingamajig.* Scholastic.

Siegler, R. S., & Ramani, G. B., (2008). Playing linear numerical board games promotes low income children's numerical development. *Developmental Science, 11*(5), 655–661.

Sullivan, P., Bobis, J., Downton, A., Hughes, S., Livy, S., McCormick, M., & Russo, J. (2020). Ways that relentless consistency and task variation contribute to teacher and student mathematics learning. In A. Coles (Ed.) *For the Learning of Mathematics Monograph 1: Proceedings of a symposium on learning in honour of Laurinda Brown* (pp. 32–37). Canada: FLM Publishing Association.

Thompson, S., De Bortoli, L., Underwood, C., & Schmid, M. (2019). *PISA 2018: Reporting Australia's Results. Volume I Student Performance.* Australian Council of Educational Research, Melbourne.

MATRIX TABLE OF THE SEQUENCES: CONCEPTS AND SKILLS

Domain	SEQUENCE	COMPARATIVE LANGUAGE	COUNTING BY ONES FORWARDS & BACKWARDS	SKIP-COUNTING	CONCEPTUAL SUBITISING	RECOGNISE 10 AS A SET/ (UNITISING)	PATTERN RECOGNITION	ADDITION & SUBTRACTION (PART–PART–WHOLE RELATIONSHIPS)	EARLY MULTIPLICATIVE THINKING	EQUIVALENCE	NUMBER PROPERTIES (COMMUTATIVE AND ASSOCIATIVE)	CONSTRUCT & INTERPRET REPRESENTATIONS	ESTIMATION	CONSTRUCT & INTERPRET REPRESENTATIONS	VISUALISATION (IMAGINE)	GENERATE A RULE	WORK SYSTEMATICALLY	EXPLAIN & JUSTIFY REASONING	FORM GENERALISATIONS
Number	Counting principles	X	X	X	X	X		X	X			X	X	X	X			X	
Number	Structure of number	X	X		X		X	X	X	X	X	X		X	X			X	X
Number	Making things equal	X	X				X	X		X	X	X		X				X	X
Number	Counting patterns		X	X	X		X		X			X		X	X	X	X	X	X
Number	Place value	X		X	X	X	X	X		X		X	X	X			X	X	X
Measurement	Informal length measuring	X	X			X						X	X	X	X	X		X	X
Measurement	Time	X		X			X	X				X	X	X	X			X	
Measurement	Informal approaches to perimeter and area	X	X	X				X	X			X	X	X	X		X	X	X
Measurement	Volume	X	X	X					X			X	X	X	X		X	X	X
Geometry	Recognising polygons											X		X	X		X	X	X
Geometry	Objects											X		X	X	X	X	X	
Geometry	Reasoning with polygons											X		X	X	X	X	X	X
Geometry	Location and transformation		X				X					X		X	X			X	X
Probability and statistics	Probability and statistics		X	X				X				X		X		X	X	X	X